Dobrivoje Popović (Ed.)

Analysis and Control of Industrial Processes

Advances in System Analysis
Editor: Dietmar P. F. Möller

Volume 1: Emil S. Bücherl (Ed.)
Proceedings of the Second World Symposium
Artificial Heart

Volume 2: Dietmar P. F. Möller (Ed.)
System Analysis of Biological Processes

Volume 3: Kiichi Tsuchiya and Mitsuo Umezu
Mechanical Simulator of the Cardiovascular System:
Design, Development and Application

Volume 4: Erhard Godehardt
Graphs as Structural Models

Volume 5: Dietmar P. F. Müller (Ed.)
System Analysis of Biomedical Processes

Volume 6: Dobrivoje Popović (Ed.)
Analysis and Control of Industrial Processes

Manuscript submitted to *Advances in System Analysis* must be original, pointing out the advancement of the contribution with respect to the actual a-priori knowledge.

Manuscript or exposé should be sent to the Editor of the Series:
Dietmar P. F. Möller, Johannes Gutenberg Universität Mainz, Physiologisches Institut, Saarstr. 21, D-6500 Mainz 1, W.-Germany.

Dobrivoje Popović (Ed.)

Analysis and Control of Industrial Processes

vieweg

Vieweg is a subsidiary company of the Bertelsmann Publishing Group International.

ISBN 978-3-528-06340-5 ISBN 978-3-322-88847-1 (eBook)
DOI 10.1007/978-3-322-88847-1

ISSN 0932-593X

Contents

Authors

Bouabdallah, N., Dipl.-Ing.	University of Bremen, Institute of Technology, Universitätsallee, 2800 Bremen 33
Breckling, B., Dr. rer. nat.	University of Bremen, Dept. of Biology (FB 2), Ecosystem-Research, Loebener Str., 2800 Bremen 33
Bruce-Boye, C., Dipl.-Ing.	University of Bremen, Institute of Automation Technology, Universitätsallee, 2800 Bremen 33
Cordes, J., Dipl.-Ing.	University of Bremen, Institute of Automation Technology, Universitätsallee, 2800 Bremen 33
Danziger, D., Dr.-Ing.	Europart GmbH, Funkschneise 5, 2800 Bremen 44
Deppner, H.-G., Dr.-Ing.	MBB, Hünefeldstr. 1–5, 2800 Bremen 1
El Zuby, O., Dr.-Ing.	R + I System GmbH, Universitätsallee 5, 2800 Bremen 33
Fedderwitz, W., Dr.-Ing.	Krupp-Atlas Elektronik, Sebaldsbrücker Heerstraße 235, 2800 Bremen 44
Heine, R., Dipl.-Ing.	University of Bremen, Institute of Automation Technology, Universitätsallee, 2800 Bremen 33
Hillert, F., Dr. oec.	University of Bremen, Institute of Automation Technology, Universitätsallee, 2800 Bremen 33
Hofmann, I., Dipl.-Ing.	University of Bremen, Institute of Automation Technology, Universitätsallee, 2800 Bremen 33
Jaeger, H., Prof. Dr. rer. nat.	University of Bremen, Institute of Applied and Physical Chemistry, Leobener Straße, 2800 Bremen 33
Kiendl, H., Prof. Dr. rer. nat.	University of Dortmund, Control Engineering Laboratory, Otto-Hahn-Str., 4600 Dortmund-Eichlinghofen
Koch, E., Dipl.-Math.	University of Bremen, Institute of Automation Technology, Universitätsallee, 2800 Bremen 33
Lorenz, I., Dipl.-Ing.	MBB-ERNO Raumfahrttechnik GmbH, Hünefeldstraße 1–5, 2800 Bremen 1
Ludyk, G., Prof. Dr.-Ing.	University of Bremen, Institute of Automation Technology, Universitätsallee, 2800 Bremen 33
Massau, C., Dipl.-Ing.	MBB-ERNO Raumfahrttechnik GmbH, Hünefeldstraße 1–5, 2800 Bremen 1
Meyer, U., Dr.-Ing.	BRESAT GmbH, Fahrenheitstr. 1, 2800 Bremen 33
Münstermann, R., Dr.-Ing.	Krupp-Atlas-Elektronik, Sebaldsbrücker Heerstraße 235, 2800 Bremen 44

Philippsen, H.-W., Dr.-Ing.	University of Bremen, Institute of Automation Technology, Universitätsallee, 2800 Bremen 33
Popovic, D., Prof. Dr.-Ing.	University of Bremen, Institute of Automation Technology, Universitätsallee, 2800 Bremen 33
Puvogel, C., Dr.-Ing.	OHB-System, Universitätsallee 27, 2800 Bremen 33
Schnare, T., Dipl.-Ing.	University of Bremen, Institute of Automation Technology, Universitätsallee, 2800 Bremen 33
Schulz-Ekloff, G., Prof. Dr. rer. nat.	University of Bremen, Institute of Applied and Physical Chemistry, Loebener Straße, 2800 Bremen 33
Tanha, A., Dipl.-Ing.	University of Bremen, Institute of Automation Technology, Universitätsallee, 2800 Bremen 33
Thiele, G., Priv.-Doz. Dr.-Ing.	University of Bremen, Institute of Automation Technology, Universitätsallee, 2800 Bremen 33
Unbehauen, H., Prof. Dr.-Ing.	Ruhr-University Bochum, Control Engineering Laboratory, Universitätsallee 50, 4630 Bochum
Wendland, E., Dipl.-Ing.	University of Bremen, Institute of Automation Technology, Universitätsallee, 2800 Bremen 33
Zimmermann, A., Dipl.-Ing.	Krupp-Atlas Elektronik, Sebaldsbrücker Heerstraße 235, 2800 Bremen 44

Preface

This book is an outcome of a 2-days Workshop on Automation Engineering held in Salzhausen (near Hamburg), F.R. Germany. The Workshop is annually organized by the Institute of Automation Technology and the Institute of Applied and Physical Chemistry of the University of Bremen. It is regularly joined by 30 to 40 participants representing the teaching and reserarch staff of both Institutes, as well as some invited colleagues from other universities and the regional industry.

Usually 15/20 selected contributions, grouped in technical sessions (system modelling and identification, control systems design, computer hardware and software, knowledge-based systems, robotics etc.), are presented at the Workshop, whereby the invited guests preferably present survey papers on current problems of modern systems engineering. For this book, 22 papers, contributed to the last two Workshops, are selected and grouped into 6 Chapters. The Chapters reflect the technical sessions listed above and give an average view on current research activities at both Institutes of the University. It is expected that the book will prove as a useful reference to the scientists and practicing engineers in the area of development and application of methods of modern systems engineering in processing industry and the robotics.

The editor wishes to thank all members of the University who have contributed to the organization of the Workshops and to the preparation of the book. His thanks particularly go to Professor G. Ludyk, W. Sobotta, G. Schulz-Ekloff, and J. Jaeger of the University of Bremen, and to Prof. H. Unbehauen (University of Bochum) and Prof. H. Kiendl (University of Dortmund) for their valuable contributions and the contributions of their scientific staff.

The editor, however, likes to point out that the book would not be possible without the essential preparatory support by Dr. G. Thiele of University of Bremen, to whom the editor would like to express the deep gratitude.

It is, finally, the pleasant obligation of the editor to thank Dr. D. P. F. Möller, the Vieweg editor of this Series, based on whose initiative the idea of editing the book was created.

Bremen, January 1990 *D. Popović*

1
Survey Papers

Review and Future of Adaptive Control Systems

H. Unbehauen

Abstract: This survey of adaptive control systems reviews the progress in theory and application. Different basic structures of adaptive control systems are described. The main design principles are sketched and it is shown that carefully designed adaptive control systems have been used successfully in a broad variety of application areas. Future perspectives in the development of adaptive control theory and requirements for practical applications will be discussed.

1. INTRODUCTION

This contribution has the aim to introduce the nonspecialist briefly to the field of adaptive control and to evaluate the actual status of this field for the more specialised control engineer. In order to have a clear guideline we start with the *definition* of adaptive control systems [1]:

> A control system is an adaptive one, if some of its properties that can be influenced are automatically adjusted according to a performance criterion in case of changes in the controlled process or it's unknown parameters.

Terms as self-adjusting, self-adapting, self-optimizing, self-tuning or auto-tuning are for this purpose regarded to be synonymous with the term "adaptive".

Adaptive control systems are a natural expansion of the classical control loop [2]. As shown in Fig. 1 an adaptive system consists of a basic control loop and a superimposed adapting system. In addition a supervisory system is superimposed upon the adaptive control system to ensure proper functioning of the subsystems and/or the whole system and to discover any failure and to induce in that case adequate reactions. The adapting system performs three characteristic functions: identification, decision and modification. The purpose of *identification* in an adaptive control system is to ascertain the properties of a system or subsystem. In the *decision* process the information acquired by the identification of the system is compared with desired properties in accordance with predetermined criteria and decided how the controller is to be adapted. The *modification* is the realization of the results of the decision process.

In adaptive control the controller settings are automatically adjusted in order to achieve good process operation over a wide range of conditions. The controller adaptation is necessary either for poorly understood processes or to compensate for unanticipated parameter changes of the process due to environmental conditions or unpredictable operating point changes. Thus adaptive control provides possibilities to control processes with uncertainties, as e.g. nonlinearities and time-varying parameters.

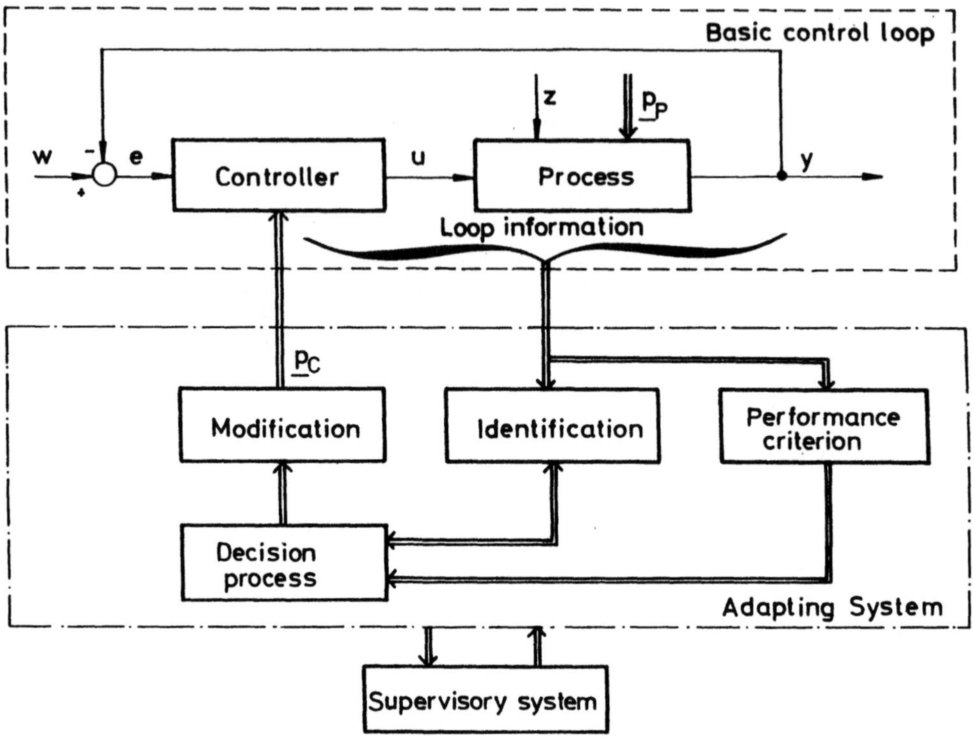

Fig. 1. Basic structure of an adaptive control system (e control deviation, u manipulating variable, w set point, y controlled variable, z disturbance, $\mathbf{p}_c$ controller parameter vector, $\mathbf{p}_p$ plant parameter vector)

Although adaptive control strategies have been discussed broadly during the last 40 years it is only in the last decade that adaptive control has found real industrial applications. This situation is based on the one hand on the progress in the development of powerful adaptive control algorithms which have reached today a mature state. On the other hand, modern microelectronics offers cheap hardware which allows an easy realization of adaptive control strategies, leading already to commercially viable solutions.

In the following sections at first the historical development of adaptive control systems will be discussed. Then a short classification and description of adaptive control principles is given. The next section shows that most adaptive schemes have nearly the same structure. This is followed by a review of recent developments in adaptive control theory as well as in practical applications. The last section is devoted to future perspectives of adaptive control systems.

2. HISTORICAL DEVELOPMENT

This section describes briefly the historical development of adaptive control systems from the beginning up to 1980. The recent developments during the current decade will be discussed later in section 5.

The adaptive control concept which provides solutions for the cases when the plant dynamics are either poorly understood or when large and unpredictable variations occur seems to be old, and first proposals and applications had already been reported during the period 1940-1950.

The period 1950-1960 is characterized by numerous trial and error approaches for the development of adaptive control systems. However, at this time the technical means for realization had either been unavailable or had been unreliable or too expensive. In one of the very early contributions to adaptive control Draper and Li [3] presented in 1951 a self-optimizing system which could automatically optimize the performance of an internal combustion engine. In 1958 Kalman [4] introduced a machine to obtain estimates of an impulse response sequence using an on-line least squares algorithm for a self-optimizing control system. In the same year a model reference adaptive control law (MRAC-systems), known as the M.I.T. rule, was proposed [5].

During the decade of 1960-1970 first attempts had been made to lay the theoretical foundations for adaptive control strategies. For the first time systematic stable design approaches based on Lyapunov's second stability method had been proposed, e.g. [6]. At the same time random disturbances had been included into the design of adaptive controllers leading to a stochastic problem formulation. The estimation of the process parameters was already solved by simple recursive least squares procedures. To handle non-stationary systems also, it was assumed that the system parameters are only slowly time-varying. The available process computers for realizing the theoretically developed adaptive control schemes at this time were very expensive and in addition unreliable so that they had been applied only to a few more prestige projects.

The beginning of the period 1970-1980 is characterized by the inception of the first self-tuning regulator (STR) of Peterka [7] and Aström and Wittenmark [8], and soon after the first implementation of the generalized self-tuning approach of Clarke and Gawthrop [9]. These controller types led to the development of new versions of numerous STR approaches, e.g. [10] including many practical applications. Ljung [11] explored the conditions for solving the important problem of the convergence of the included parameter estimation. Another important milestone was the introduction of Monopoli's [12] augmented error approach to the design of stable MRAC-systems, which provided the design of globally asymptotically stable adaptive control algorithms and which can be considered as basis of several important contributions not only for a resolution of the longstanding problem of stability [13 to 16], but also a unification of different approaches to adaptive control and a realization of the equivalence of MRAC- and STR-systems [17, 18]. A very similar approach for the design of stable MRAC-systems was proposed by Landau [19] using the concept of Popov's hyperstability criteria. During this period the introduction of the microprocessor provided a great progress for adaptive control, especially in may practical realizations on a broad scale.

3. CLASSIFICATION AND DESCRIPTION OF BASIC STRUCTURES

According to Fig. 2 there are various possibilities to classify adaptive control systems. A

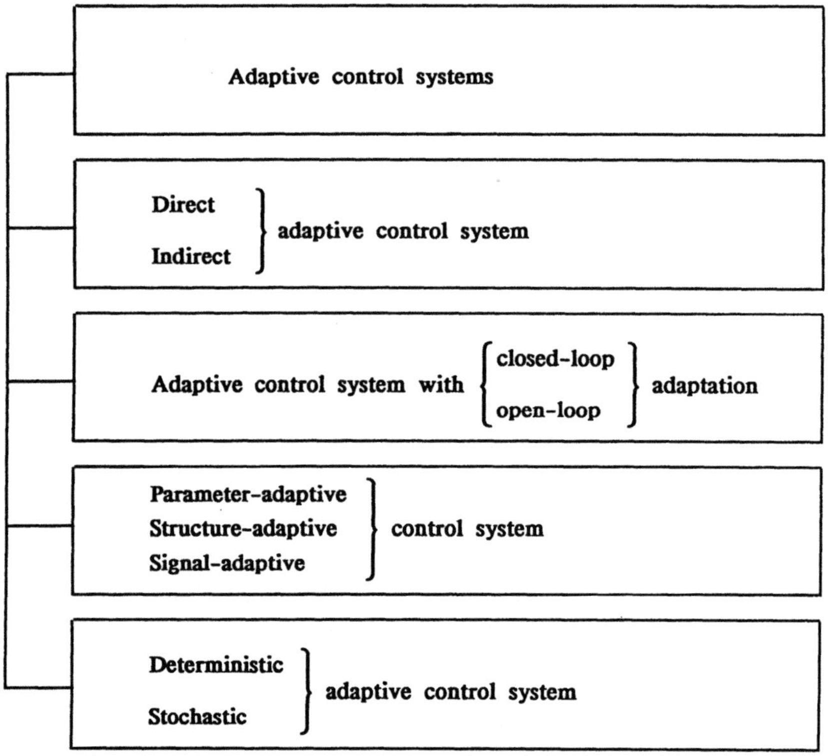

Fig. 2. Classification of adaptive control systems

control system is to be called *direct* adaptive control system if the decision process is performed without explicitly generating a model of the controlled process, otherwise it is called *indirect* adaptive control system. It should be mentioned that without regarding whether there is a direct (sometimes also called: implicit) or indirect (sometimes also called: explicit) adaptive control scheme it is possible to describe different principles and ways of functioning of adaptive controllers. So Fig. 3 shows different direct and

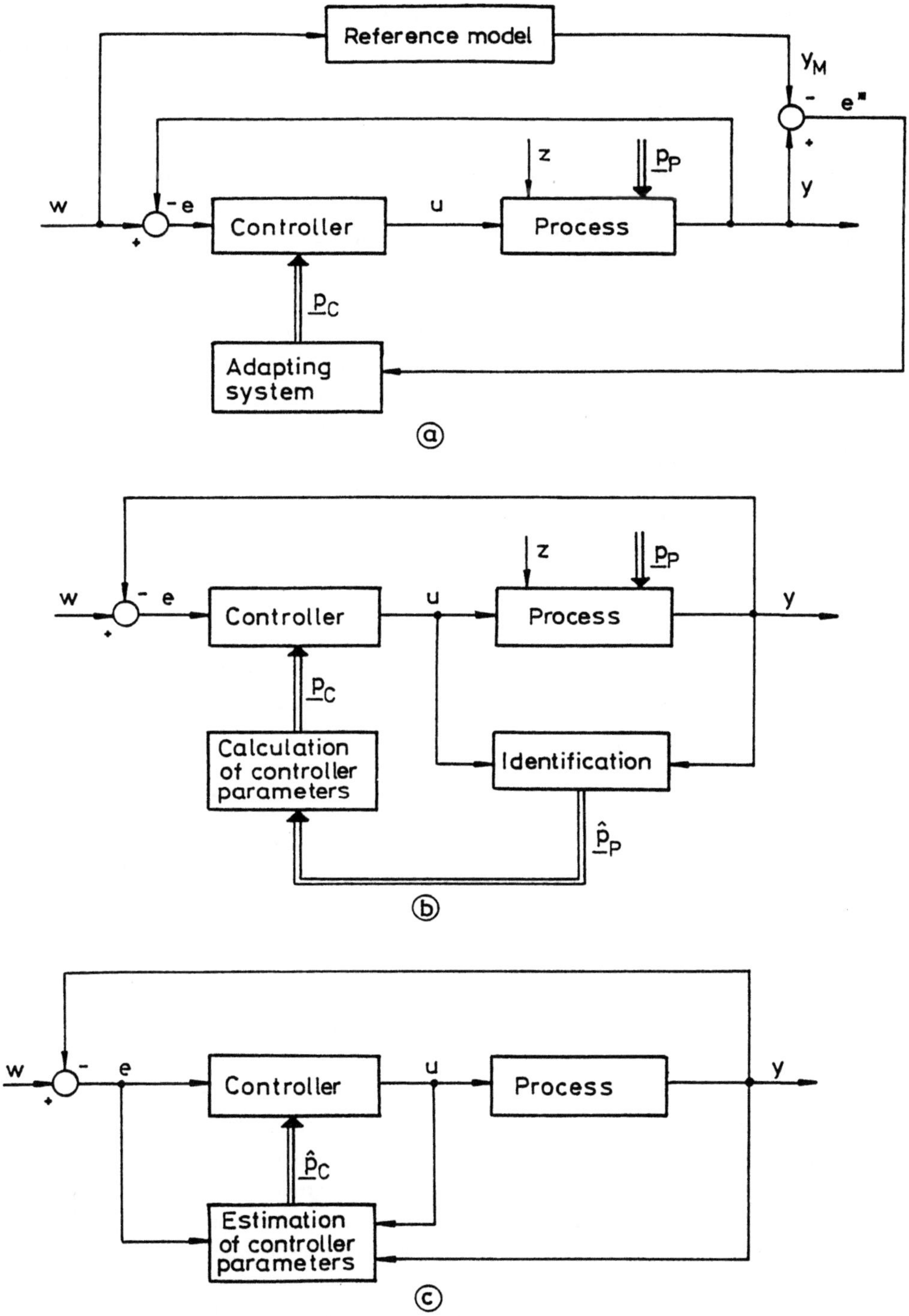

Fig. 3. Direct and indirect adaptive control structures: (a) and (c) represent direct structures, whereas (b) shows an indirect structure

indirect adaptive control systems. Fig. 3a represents an adaptive controller according to the model reference adaptive controller (MRAC-) principle. This is a direct parameter-adaptive scheme with closed loop adaptation for minimizing the output error e*. Fig. 3b shows an adaptive controller without a reference model, but the identification usually includes an identification model. Hence we have an indirect parameter-adaptive controller with closed-loop adaptation. Fig. 3c represents a direct parameter-adaptive control system with closed-loop adaptation. Here the controller parameters are directly estimated. The direct and indirect self-tuning controller (STC) belong to structures represented in Figures 3b and 3c.

All structures of Fig. 3 are adaptive control systems with *closed-loop adaptation*, because the result of the adaptation is continuously compared with the preset performance criteria in a closed-loop. If there is no such continuous comparison and no feedback of the results of the adaptation to the decision process, but instead a modification of the adaptive control system in accordance with preset schedules it is an adaptive control system with *open-loop adaptation*. The classical procedure of parameter scheduling (often denoted as gain scheduling) would be contained in this class of adaptive control systems. The principle structure of this type of adaptive control is given in Fig. 4.

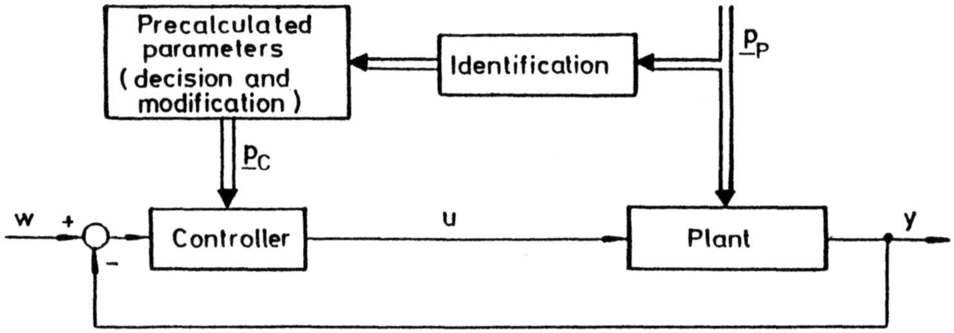

Fig. 4. Principle of gain or parameter scheduling controller

An adaptive control system will be called *parameter-*, *structure-*, *signal-adaptive* depending upon whether for the modification of the basic control loop parameters of the controller, the structure of the controller or a signal acting upon the control loop is changed. In the sense of this definition is e.g. an extremal-value control system a special type of signal-adaptive control system in which the reference signal is adapted. Fig. 5 shows examples of a structure- and a signal-adaptive control system.

A *deterministic (stochastic)* control system is one in which the decision process is based upon a deterministic (stochastic) performance index. In view of these definitions structuring of adaptive control systems according to their relevant properties as shown in

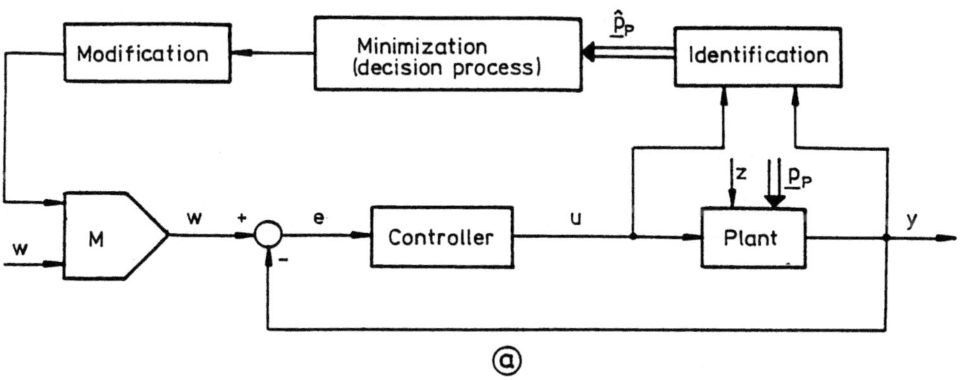

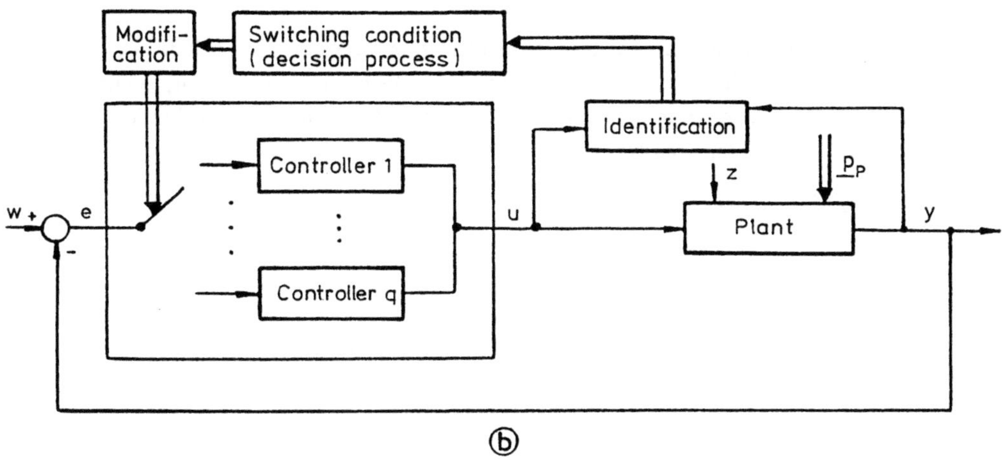

Fig. 5. (a) Extremal value controller: Signal-adaptive control system with closed-loop adaptation; (b) structure-variable adaptive control system (direct of indirect) with closed-loop adaptation.

Fig. 2 is sufficient. Any adaptive control system can contain more than one of these properties. Thus, as example, a classical self-tuning controller (STC) using minimum variance principle would, according to these definitions, be an indirect stochastic, parameter-adaptive control system with closed loop adaptation. An "adaptive controller with model-based adaptation" would be a direct, deterministic, parameter-adaptive control system with closed-loop adaptation.

4. DESIGN PRINCIPLES

The design principles mentioned in the following will be described briefly. As adaptive control is based on simultaneous explicit or implicit process identification and control, the problems of on-line parameter estimation must be dealt with primarily. The process identification and the adaptation mechanisms are usually both realized by digital process computers. Therefore, the corresponding systems are described in discrete-time form.

4.1. Recursive process identificaiton

Most adaptive control algorithms are based on a linearized process model, which, for a typical single-input/single-output (SISO) system is given by the linear difference equation

$$y_M(k) = - \sum_{\nu=1}^{n} a_\nu y_M(k-\nu) + \sum_{\nu=0}^{n} b_\nu u(k-\nu). \tag{1}$$

For a realistic description of the process model it is necessary to include an additional disturbance model as in Fig. 6, where $r_M(k)$ is a stochastic noise signal, which can be thought to be generated from a white noise signal ε (with normal distribution and zero mean) by the noise filter transfer function

$$G_r(z) = \frac{1}{A(z^{-1})} G_r^*(z). \tag{2}$$

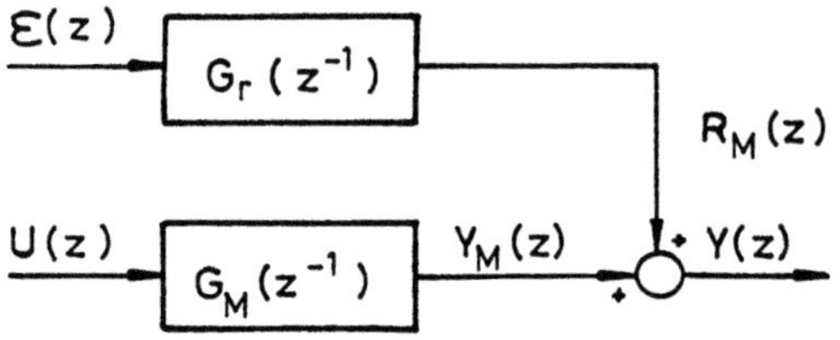

Fig. 6. Complete model structure of the process

It follows using z-transformation from Fig. 6 that

$$Y(z) = Y_M(z) + G_r(z)\ \varepsilon(z). \tag{3}$$

By inserting Eqs. (1) and (2) into Eq. (3) and multiplying by $A(z^{-1})$ we obtain

$$A(z^{-1})Y(z) - B(z^{-1})U(z) = G_r^*(z)\ \varepsilon(z) = V(z), \tag{4}$$

where V(z) is defined as general model error and

$$A(z^{-1}) = 1 + a_1 z^{-1} + \ldots + a_n z^{-n} \tag{5}$$

$$B(z^{-1}) = b_0 + b_1 z^{-1} + \ldots + b_n z^{-n}. \tag{6}$$

Eq. 4 defines an ARMAX-model. Depending on the selection of $G_r^*(z)$ all usual model structures are described by this equation [2]. E.g. the selection of $G_r^*(z) = 1$ leads to the least squares (LS) techniques, which will be used for simplicity but without loss of generality in the following.

Introducing the *data vector*

$$\mathbf{m}(k) = [-y(k-1) \; \dots \; -y(k-n) \, | \, u(k-1) \; \dots \; u(k-n)]^T, \tag{7}$$

and the *parameter vector* for the plant $\mathbf{p}_M \equiv \mathbf{p}$, i.e.

$$\mathbf{p} = [a_1 \; \dots \; a_n \, | \, b_1 \; \dots \; b_n]^T \tag{8}$$

under the assumption $b_0=0$ (which usually is fulfilled for physical systems) the output signal is obtained directly from Eq. (4) as

$$y(k) = \mathbf{m}^T(k) \; \mathbf{p} + \varepsilon(k). \tag{9}$$

The parameter estimation problem is to find an estimation $\hat{\mathbf{p}}$ of $\mathbf{p}$ using the known data vector $\mathbf{m}(k)$ such that the loss function for N measurements

$$I = I(\mathbf{p}) = \frac{1}{2} \sum_{k=n+1}^{n+N} \varepsilon^2(k) = \frac{1}{2} \boldsymbol{\varepsilon}^T(N) \; \boldsymbol{\varepsilon}(N) \overset{!}{=} \text{Min} \tag{10}$$

becomes minimal. The solution of this minimization problem can be obtained directly by inserting Eq. (9) into Eq. (10) by collecting N pairs of measurements and batch-wise data processing. In adaptive systems, the recursive solution of this problem, however, is prefered.

The recursive estimation of the LS-model is given by the following equations:

$$\hat{\mathbf{p}}(k+1) = \hat{\mathbf{p}}(k) + \mathbf{q}(k+1) \; \hat{\varepsilon}(k+1) \tag{11}$$

$$\mathbf{q}(k+1) = \mathbf{P}(k)\mathbf{m}(k+1)[1+\mathbf{m}^T(k+1)\mathbf{P}(k)\mathbf{m}(k+1)]^{-1} \tag{12}$$

$$\mathbf{P}(k+1) = \mathbf{P}(k) - \mathbf{q}(k+1)\mathbf{m}^T(k+1)\mathbf{P}(k) \tag{13}$$

$$\hat{\varepsilon}(k+1) = y(k+1) - \mathbf{m}^T(k+1)\hat{\mathbf{p}}(k). \tag{14}$$

For the application of this estimation algorithm a suitable choice of the initial values $\hat{\mathbf{p}}(0)$ and $\mathbf{P}(0)$ must be made. While the choice of $\hat{\mathbf{p}}(0)$ is not critical, $\mathbf{P}(0)$ should be selected as a diagonal matrix with large elements, which will cause rapid changes of $\hat{\mathbf{p}}(k)$ at the beginning. During the calculation the values of the diagonal elements are reduced so that $\hat{\mathbf{p}}(k)$ changes only slowly. This may lead to convergence of parameters. On the other hand for slowly varying process parameters and for large values of k the algorithm may become sluggish. This can be circumvented e.g. by introducing a variety of special techniques, for example by introducing a variable weighting factor $1/\lambda(k)$ to the matrix $\mathbf{P}(k)$ or using special resetting techniques [20].

4.2. Controller design

The design of a single-input/single-output adaptive controller can, in general, be formulated by two laws:

a) The *control law*

$$u(t) = -\mathbf{p}_c^T(t)\mathbf{m}_c(t), \tag{15a}$$

or

$$u(k) = -\mathbf{p}_c^T(k)\mathbf{m}_c(t), \tag{15b}$$

where $\mathbf{p}_c$, in the case of a direct adaptive controller structure, is directly estimated ($\mathbf{p}_c \equiv \hat{\mathbf{p}}_c$) without an explicitely generated plant model, whereas in the case of an indirect adaptive controller structure $\mathbf{p}_c$ is calculated on the basis of estimated plant parameters $\hat{\mathbf{p}}_p$. The measuring vector $\mathbf{m}_c$ contains all information about the signals in the loop. In order to obtain a causal control law which is linear in the parameters it is eventually necessary to introduce filters for filtering the signals within the loop, including in the case of MRAC-systems the model error e^* and to augment the model error by adding auxiliary signals. Thus e^* becomes ε which is defined as an *"augmented" error signal*. To ensure stability the vector $\mathbf{m}_c$, which in general contains functions of the process input and output signals, must be generated to ensure boundedness of u and y and asymptotic convergence of ε.

b) The *adaptive law* is in general given by

$$\dot{\mathbf{p}}_c(t) = -\mathbf{f}_1(t) \tag{16a}$$

or

$$\mathbf{p}_c(k) = \mathbf{p}_c(k-1) - \mathbf{f}_1(k). \tag{16b}$$

In the case that no reference model is used, as e.g. in self-tuning controller structures, $\mathbf{p}_c$ is, according to the selected design procedure, either calculated on the basis of estimated plant parameters according to Eqs. (11) to (14) (indirect approach) or directly estimated (direct approach). In the latter case we have

$$\hat{\mathbf{p}}_c(k) = \hat{\mathbf{p}}_c(k-1) + \mathbf{P}_c(k-1)\rho_c(k)\hat{\varepsilon}_c(k), \tag{16c}$$

where $\rho_c(k)$ is a regression vector obtained from sensed signals within the control loop and $\hat{\varepsilon}_c(k)$ is the prediction error. For a recursive LS-estimator both these variables and the matrix $\mathbf{P}_c$ can be obtained in accordance to Eqs. (11) to (14).

The design of a MRAC-system is usually based either on Lyapunov's stability theory or Popov's hyperstability method, both with state feedback or output feedback [2]. The main idea in applying stability theory to MRAC-systems is to transform the highly nonlinear "error system" as shown in Fig. 7, where the model error e^* either in

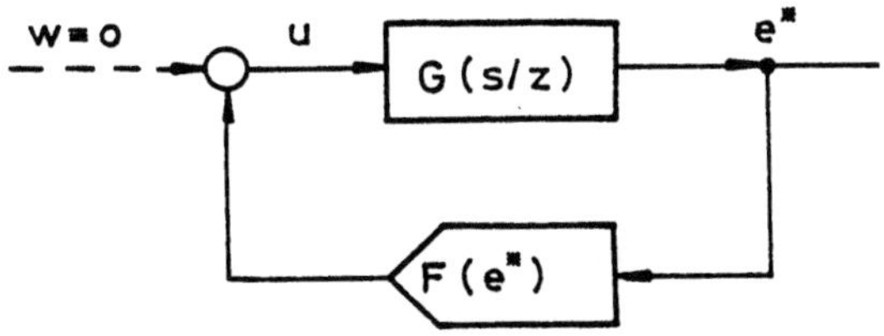

Fig. 7. Nonlinear standard "error system"

continuous or discrete form represents the output signal of a linear time-invariant system, described by the transfer function G, while F contains all nonlinear and time-variant subsystems. This structure has the advantage that the stability of the overall MRAC-system can be obtained from the individual properties of the linear and nonlinear subsystems, e.g. the linear subsystem must be strictly positive real (s.p.r.). Thus the output error

$$e^* = c^T \tilde{e}^* \tag{17}$$

or the elements $\tilde{e}^*_\ell$ of the state error vector

$$\tilde{e}^* = [\tilde{e}^*_1 \dots \tilde{e}^*_\ell \dots \tilde{e}^*_n]^T \tag{18}$$

can in general be represented by a nonlinear time-varying differential equation of the form

$$\dot{\varepsilon}(t) = f_2[\varepsilon(t),\ p_c(t),\ t] \tag{19a}$$

or a difference equation of the form

$$\varepsilon(k+1) = f_2[\varepsilon(k),\ p_c(k),\ k], \tag{19b}$$

where ε can be replaced either by e^* or $\tilde{e}^*$, and p_c is the parameter vector or the corresponding parameter error vector of the adaptive controller, for which the adaptation law according to Eq. (16a,b) must be developed, such that all signals are uniformly bounded and

$$\lim_{t\to\infty} \varepsilon(t) = 0 \tag{20}$$

using all available data. The function f_1 in Eq. (16) can be obtained either by estimation or from filtered process measurements. f_1 includes a multiplicative connection between the error ε and some regression vector ρ which is in general represented by functions of u and y [2].

This discussion shows that the basic structures of adaptive control systems, either MRAC-schemes or STC-schemes, are identical although the background of MRAC was the servo control problem, whereas the STC originally had been designed for the stochastic regulation problem. Both principles are characterized by two feedback loops. However, the design principle of these two loops is different.

5. RECENT DEVELOPMENTS IN THEORY AND APPLICATIONS

Since 1980 a number of surveys [21 to 28] and books [2; 29 to 44] as well as conference proceedings [45 to 60] and special issues of important technical journals [61, 62] had been published and a special new technical journal [63] has been edited, which provide exhaustive information on the status and application of adaptive control systems for the interested reader.

5.1. Developments in theory

During the first half of the current decade 1980-1990 the proof of global stability of adaptive systems and convergence of the included parameter estimation under restrictive assumptions had been well developed. The extension of single-input/single-output adaptive control sytems to multivariable structures took place [64, 65]. Further theoretical developments have been concerned with several different aspects of adaptive control structures, which, however, in many cases represent combinations or modifications of already known adaptation algorithms. Other problems such as adaptive sampling, multi-loop cascaded structures and hybrid adaptive control structures as well as special design schemes based on quadratic cost functions and pole placement have been reported. First attempts have also been made to develop adaptive control systems for time-varying and nonlinear processes. Proposals have been made how to deal with non-minimum phase plants in the design of adaptive control systems [66 to 68]. Details of this development are reported in [24, 25].

Globally stable adaptive control systems have been developed for plants which can be described by a linear time-invariant, minimum-phase model of known structure. In addition the allowable disturbances has to be mean Gaussian random signals. However, stability will no more be preserved when the plant is subject to unknown bounded disturbances or unmodelled dynamics [69]. This led to the problem of *robustness* of adaptive control systems. Bounded unknown disturbances and unmodelled higher dynamics can then make the basic adaptive scheme unstable. The major developments in adaptive control theory since about 1985 have been devoted to overcome these problems. Two main approaches have been proposed for the solution of these problems:

a) The concept of *persistent excitation* [70] is directly related to the speed of convergence of the controller parameters to their final values, as well as the bounds on the magnitude of the parameter errors which again is connected with the uniform asymptotic stability of the adaptive system. Hence appropriate persistent excitation of the reference input is necessary to ensure that some of the internal signals become dominantely rich.

b) The concept of *robust parameter estimation* had been included into adaptive control systems. Robustness of adaptive control systems means the preservation of stability or boundness properties when ideal conditions are no more met. To prevent instability the parameter drift caused by bounded external disturbances can be concentrated through eliminating the integral action of the adaptive control law. Therefore, several proposals have been made to modify the adaptation laws:

- Application of a *dead zone* [71] into the adaption law: If an upper limit of the disturbances is known, then the adaptation is switched off when the tracking error is between a specified threshold.

- Introduction of the so called σ-modification [72] into the adaptation law in order to obtain stability of the adaptive system under the influence of limited disturbance magnitude. Within this approach the usual integral parameter adaptation law, e.g. according to Eq. (16),

  $$\dot{\mathbf{p}}(t) = -\mathbf{f}_1(t) \tag{21a}$$

  or

  $$\mathbf{p}(k) = \mathbf{p}(k-1) - \mathbf{f}_1(k) \tag{21b}$$

  is replaced by

  $$\dot{\mathbf{p}}(t) = -\sigma\mathbf{p}(t) - \mathbf{f}_1(t) \qquad \sigma > 0 \tag{22a}$$

  $$\mathbf{p}(k) = \sigma\mathbf{p}(k-1) - \mathbf{f}_1(k) \qquad |\sigma| < 1. \tag{22b}$$

  This modification should be applied only if the norm of the parameter vector exceeds some a priori defined value $\|\mathbf{p}\| > N_0$. In this case, simple stability is guaranteed for stochastic disturbances and errors for unmodelled high-frequency modes of the process. Thus the stability of adaptive control sytems using reduced-order models is obtained.

- The e-modification [73] is very similar to the σ-modification scheme, where only σ in Eq. (22) is replaced by a term proportional to $|e|$, where e represents the output error ($e=\hat{y}-y$). The corresponding correction term in the adaptation law tends to zero with the output error. Hence if no disturbance is present the mismatch between the plant and the model vanishes.

By these modifications parameter drift caused by bounded disturbances can be prevented, however, the important problem of unmodelled dynamics cannot be dealt with directly, because the corresponding disturbances are no more in general unbounded. To solve this problem the concept of data normalization has been introduced, which transfers the problem of unbounded disturbances to one of bounded disturbances [20, 71, 74]. Thus it can be established that such a modified algorithm guarantees global stability and convergence of an adaptive control system with a dead-zone in the presence of unmodelled dynamics.

During the last decade adaptive long-range predictive control systems have received considerable attention in theoretical treatment. These types of adaptive controllers are especially appropriate to overcome such challenging control problems as unknown and/or variable dead-time, non-minimum phase behaviour and plant-model mismatch [28, 75]. Many successful industrial applications have been reported.

Other interesting contributions during the recent years have been devoted to the problems of chaotic dynamics in adaptive systems [76, 77] and to decentralized adaptive control systems [78]. Another shift of interest is directed towards to techniques which are related to the new field of H_∞-stability [79]. Controllers which preserve this characteristic within a closed loop can never become instable by unmodelled dynamics. The realisation of explicit LQG-self-tuning regulators seems to be possible, however until recently the computational load involved was still very high [80]. Other developments are concerned with the application of adaptive control to nonlinear dynamics processes [81] and to more intelligent adaptive control schemes [82].

5.2. Application of adaptive control

Even though there had been many practical realizations of adaptive control schemes during the past 40 years, see e.g. [24, 83, 84] it is only since 1980 that adaptive techniques became of real interest in industrial applications. Whereas the first implementations on a broader scale had been performed during the 1970s using minicomputers the number of applications increased rapidly with the availability of cheap and powerful microprocessors since about 1980. Besides special realization or more generally applicable *programme packages* [85] a considerable number of *standard industrial adaptive controllers* is today commercially available [86 to 88].

Adaptive techniques are involved both in single-loop controllers and as standard elements of large process control systems. For special applications adaptive control schemes are often tailored. More than 100,000 adaptive loops are today in industrial operation. The industrial solutions are usually fitted with sophisticated safety networks or supervisory logic units. The man-machine interface must allow an easy operation, whereby only systems with a minimal number of adjustments of controller parameters are accepted by the plant operators. Hence adaptive PID-type controllers have good perspectives for industrial acceptance. Whereas the first generations of adaptive controllers had been based on 8-bit microprocessors, the second generation of advanced adaptive control systems applies the new 32-bit processors.

Successful and broad applications are registered from the fields of chemical industry, steel industry, power systems, electromechanical and hydraulic drives, airborne vehicles, ship control, industrial robots, automotive systems, pulp and paper industry, cement industry, heating and air-conditioning systems, satellite altitude control, medical and biotechnical sytems.

This very brief discussion shows the suprisingly broad area of applications of adaptive control systems. Obviously more heuristic ad hoc solutions are becoming rare, whereas most applications are based today on well established approaches of modern adaptive control theory. In special cases, however, the theoretical standard approaches have to be slightly modified to overcome special demands of the problem.

It should also be mentioned that modern adaptive control schemes can not yet be applied routinely by an inexperienced engineer. A lot of design specifications including as much as possible "a-priori"-knowledge about the process must still be regarded. In addition, various practical aspects for implementation of adaptive control schemes, including e.g. robustness, signal conditioning, parameter tracking, estimator wind-up, reset action, start-up etc. have to be considered by the user.

6. FUTURE PERSPECTIVES

Because of the rapidly developing field of adaptive control systems it appears to be difficult to give some more general future perspectives. However, it is possible to identify certain trends in necessary developments, both in theory and industrial applications.

Obviously some trends for future research in the theoretical development of adaptive control systems are the following:

- development of new adaptive controllers for strongly nonlinear and fast time-varying stochastic plants,
- improvement of decentralized adaptive control schemes,
- new adaptive controllers for strongly coupled multivariable plants,
- implementation of on line-structure detection algorithms into strongly nonlinear adaptive control schemes,
- introduction of more intelligence (especially learning) into the operation of adaptive control systems,
- exploration of the area of adaptive control of distributed parameter systems,
- development of new sophisticated extremal value controllers,
- improvement of methods for convergence and global stability,
- introduction of expert systems for control, supervision and adaptation,
- improvement of on-line identification of linear and nonlinear processes,
- implementation of fast parallel data processing in adaptive algorithms,
- introduction of the concept of H_∞ into adaptive control schemes.

The trends in industrial applications are obviously directed towards more and new applications. Probably the automotive area will, in the future, be one of the most promising, especially for standard solutions. But also in process industries as well as in power systems a great variety of potential applications can be expected. For these applications new and even more powerful adaptive controller schemes have to be provided. The next generation of advanced adaptive control systems will no more only have simple PID-like control algorithms. However, the sophisticated adaptive control schemes must be both easily applicable and understandable for the operating personell. In addition a good man-machine interface must be available.

Adaptive control theory has reached today a high degree of maturity. A lot of powerful design methods are available now for the experienced control engineer, including also computer-aided design packages for adaptive controllers. In addition standard industrial adaptive controllers can be applied routinely already in many practical cases. The numerous applications of adaptive control systems in a broad area of technical fields, discussed in the previous section, indicate that adaptive control can be successfully used in many situations. However, a number of unsolved challenging problems is still waiting for engineering solutions. For achieving this, further efforts both in practice and theory have to be undertaken.

REFERENCES

[1] VDI/VDE-Richtlinien 3685/Part 1: Properties of adaptive control instruments. Beuth-Verlag, Berlin 1990 (in press).

[2] Unbehauen, H.: Regelungstechnik III: Identifikation, Adaption, Optimierung. 3. Auflage, Vieweg-Verlag, Braunschweig 1988.

[3] Draper, C. and Y. Li: Principles of optimalizing control systems and an application to the internal combustion engine. ASME-Publication, New York 1951.

[4] Kalman, R.: Design of self-optimising control systems. Trans. ASME **80** (1958), No. 2, pp. 468-478.

[5] Whitaker, H. et al.: Design of a model reference adaptive control system for aircraft. Report R-164 Instrumentation Lab., M.I.T. Boston 1958.

[6] Shackclosh, B. and R. Butchart: Synthesis of model reference adaptive control system by Lyapunov's second method. Proceed. IFAC-Symposium, Teddington (1965), pp. 145-152.

[7] Peterka, V.: Adaptive digital regulation of noisy systems. Proceed. of IFAC-Symposium, Prag 1970, 6.2, pp. 1-11

[8] Aström, K. and B. Wittenmark: On self-tuning regulators. Automatica **9** (1973), pp. 185-199.

[9] Clark, D. and P. Gawthrop: A self-tuning controller. Proceed. IEE 122 (1975), pp. 929-934.

[10] Wellstead, P. et al.: Pole assignment self-tuning regulators. Proceed. IEE 126 (1979), pp. 781-787.

[11] Ljung, L.: On positive real transfer functions and the convergence of some recursive schemes. IEEE Trans. AC 22 (1977), pp. 539-551.

[12] Monopoli, R.: Model reference adaptive control with an augmented error signal. IEEE Trans. AC 19 (1974), pp. 474-484.

[13] Narendra, K. and Y. Lin: Stable discrete adaptive control. IEEE Trans. AC 25 (1980), pp. 456-461.

[14] Schmidt, Chr.: Ein Beitrag zur Realisierung adaptiver Regelungssysteme mit dem Prozeßrechner. Diss. Ruhr-Universität, Bochum 1979.

[15] Morse, A.: A global stability of parameter-adaptive control systems. IEEE Trans. AC 25 (1980), pp. 433-439.

[16] Goodwin, G. et al.: Discrete-time multivariable adaptive control. IEEE Trans. AC 25 (1980), pp. 449-456.

[17] Egardt, B.: Unification of some continuous-time adaptive control schemes. IEEE Trans. AC 24 (1979), pp. 588-592.

[18] Ljung, L. and I. Landau: Model reference adaptive systems and self-tuning regulators - some connections. Proceed. 7th IFAC World Congress, Helsinki 1978, pp. 1973-1979.

[19] Landau, I.: Adaptive control. M. Dekker Inc., New York 1979.

[20] Shah, L. and W. Cluett: RLS based estimation schemes for self-tuning control. In Warwick, K. (Ed.): Implementation of self-tuning controllers. P. Peregrinus, London 1988, pp. 23-63.

[21] Isermann, R.: Parameter adaptive control algorithms - a tutorial. Automatica 18 (1982), pp. 513-528.

[22] Aström, K.: Theory and applications of adaptive control - a survey. Automatica 19 (1983), pp. 471-486.

[23] Voronov, A. et al.: State-of-the-art and prospects of adaptive sytems. Automatica 20 (1984), pp. 547-557.

[24] Unbehauen, H.: Theory and application of adaptive control. Proceed. IFAC-conference on digital computer applications to process control. Vienna 1985, pp. 3-19.

[25] Unbehauen, H.: Stand der Theorie und Anwendung adaptiver Regelverfahren. Proceed. SGA-Tagung, Zürich 1985, pp. 17-42.

[26] Isermann, R.: Stand und Entwicklungstendenzen bei adaptiven Regelungen. Automatisierungstechnik at 35 (1987), pp. 133-143.

[27] Seborg, D. et al.: Adaptive control strategies for process control: A survey. AIChE Journal 32 (1986), pp. 881-913.

[28] Garcia, C. et al.: Model predictive control: Theory and practice - a survey. Automatica 25 (1989), pp. 335-348.

[29] Unbehauen, H. (Ed.): Methods and applications in adaptive control. Springer-Verlag, Berlin 1980.

[30] Narendra, K. and R. Monopoli (Ed.): Applications of adaptive control. Academic Press, New York 1980.

[31] Harris, C. and S. Billings (Ed.): Self-tuning and adaptive control. Theory and applications. P. Peregrinus Ltd., London 1981.

[32] Ioannou, P. and P. Kokotovic: Adaptive systems with reduced models. Springer-Verlag, Berlin 1983.

[33] Goodwin, G. and K. Sin: Adaptive filtering, prediction and control. Prentice-Hall, Englewood Cliffs (New York) 1984.

[34] Narendra, K. (Ed.): Adaptive and learning systems. Plenum Press, New York 1986.

[35] Landau, I. and L. Dugard: Commande adaptive: aspects pratiques et théoriques. Masson, Paris 1986.

[36] Böcker, J., Hartmann, I. and Ch. Zwanzig: Nichtlineare und adaptive Regelungssysteme. Springer-Verlag, Berlin 1986.

[37] Anderson, B. et al.: Stability of adaptive systems. MIT Press, Cambridge (USA) 1986.

[38] Chalam, V.: Adaptive control systems. M. Dekker, New York 1987.

[39] Gawthrop, P.: Continuous-time self-tuning control (Vol. I). J. Wiley, New York 1987.

[40] Schulze, K. and K. Rehberg: Entwurf von adaptiven Regelsystemen. VEB Verlag Technik, Berlin 1988.

[41] Warwick, K. (Ed.): Implementation of self-tuning controllers. P. Perigrinus Ltd., London 1988.

[42] Yadykin, I.B. et al.: Adaptive control of continuous technical processes (in Russian). Eergoatomizdat, Moskau 1985.

[43] Aström, K. and B. Wittenmark: Adaptive control. Addison-Wesley, Reading (USA) 1989.

[44] Narendra, K. and A. Annaswamy: Stable adaptive systems. Prentice Hall, Englewood Cliffs (USA) 1989.

[45] Narendra, K. (Ed.): Proceed. of the Workshop on application of adaptive system theory, Yale University. New Haven, May 1981.

[46] Bekey, G. and G. Saridis (Ed.): Preprints of 6th IFAC Symposium on Identification and system parameter estimation, Arlington. McGregor & Weiner, Washington 1982.

[47] Mosca, E. (Ed.): Proceed. of the ISIS Workshop adaptive control, Florence, Institute of Systems and Informatics. University of Florence 1982.

[48] N.N.: Prepr. IEEE Conference on application of adaptive and multivariable control, Hull. Cotsword Press, Oxford 1982.

[49] N.N.: Preprints of IFAC Workshop on adaptive systems in control and signal processing, San Francisco 1983.

[50] Narendra, K. (Ed.): Proceed. of the 3rd Workshop on applications of adaptive sytem theory, Yale University. New Haven, 1983.

[51] Barker, H. and P. Young (Ed.): Preprints of the 7th IFAC/IFORS Symposium on identification and system parameter estimation, York. Pergamon Press, Oxford 1985.

[52] Schaufelberger, W. (Ed.): 59. SGA-Tagung Nichtlineare, adaptive und robuste Regelung. ETH Zürich, 1985.

[53] Unbehauen, H. (Ed.): Proceed. of IFAC Workshop on adaptive control of chemical processes, Frankfurt/Main 1985, Pergamon Press, Oxford 1986.

[54] Aström, K. and B. Wittenmark (Ed.): Preprints of 2nd IFAC Workshop on adaptive systems in control and signal processing, Lund. Studentliteratur AB, Lund 1986.

[55] Narendra, K. (Ed.): Proceed. of the 5th Yale Workshop on applications of adaptive systems theory, Yale University. New Haven, 1987.

[56] Chen, H. (Ed.): Preprints of 8th IFAC/IFORS-Symposium on identification and system parameter estimation, Beijing. Pergamon Press, Oxford 1988.

[57] Kümmel, M. (Ed.): Preprints of IFAC International symposium Adchem '88 on adaptive control of chemical processes, Copenhagen, Pergamon Press, Oxford 1988.

[58] Goodwin, G. (Ed.): Preprints of IFAC Workshop on robust adaptive control, Newcastle. Institution of Engineers, Australia, Barton, A.C.T., 1988.

[59] Gawthrop, P. (Ed.): International conference control 88, Oxford. Conference Publication No. 285, Chameleon Press, Wandsworth, London 1988.

[60] Owens, D., T. Durrani and M. Grimble (Ed.): Preprints of IFAC Symposium ACASP 89 on adaptive control and signal processing, Glasgow, Imediaprint, London, 1989.

[61] Optimal control, applications & methods (OCAM), John Wiley & Sons: Special issue on self-tuning control, 3 (1982).

[62] Automatica (The IFAC Journal, Pegamon Press, Oxford): Special issue on adaptive control, **20** (1984).

[63] Grimble, M.J. (Ed.): Adaptive Control and Signal Processing, John Wiley & Sons.

[64] Elliot, H.: An overview of multivariable adaptive control. Proceed. 3rd Yale Workshop on application of adaptive system theory, New Haven 1983, pp. 211-219.

[65] Hahn, V.: Direkte adaptive Regelstrategien für die diskrete Regelung von Mehrgrößensystemen. Diss. Ruhr-Universität, Bochum 1983.

[66] Hahn, V. and H. Unbehauen: Direct adaptive control schemes for nonminimum phase systems. Preprints IEEE Conference on application of adaptive and multivariable control, Hull (UK), 1982, pp. 170-175.

[67] Nöth, G.: Self-tuning-Strategien zur Regelung nichtminimalphasiger Regelstrecken. Diss. Ruhr-Universität Bochum, 1982.

[68] Unbehauen, H.: Entwurf und Realisierung neuer adaptiver Regler nach dem Modellvergleichsverfahren. Automatisierungstechnik at **37** (1989), pp. 249-257.

[69] Rohrs, C. et al.: Robustness of adaptive control algorithms in the presence of unmodelled dynamics. Proceed. 21st Conference on Decision and Control (CDC), Orlando (USA) 1982, pp. 3-11. See also: IEEE Trans. AC **30** (1985), pp. 881-889.

[70] Narendra, K. and A. Annaswamy: Persistent excitation of adaptive systems. Int. J. of Control **45** (1987), pp. 127-160.

[71] Egardt, B.: Stability of adaptive controllers. Springer-Verlag, Berlin 1979.

[72] Ioannu, P. and P. Kokotovic: Adaptive systems with reduced models. Springer-Verlag, New York 1983.

[73] Narendra, K. and A. Annaswamy: A new adaptive law for robust adaptive control without persistent excitation. IEEE Trans. AC **32** (1987), pp. 134-145.

[74] Praly, L.: Global stability of a direct adaptive control scheme with respect to a graph topology. In: Narendra, K. (Ed.): Adaptive and learning systems - Theory and application. Plenum Press, New York 1986.

[75] Krämer, K. and H. Unbehauen: Survey to adaptive long-range predictive control. Proceed. 12th IMACS World Congress, Paris 1988, pp. 358-363.

[76] Anderson, B.: Adaptive systems, lack of persistency of excitation and bursting phenomena. Automatica **21** (1985), pp. 247-258.

[77] Ydstie, B. and M. Golden: Chaotic dynamics in adaptive systems. Prepr. IFAC Workshop on robust adaptive control, Newcastle (Australia) 1988, pp. 14-19.

[78] Wiemer, P. and H. Unbehauen: Decentralised discrete model reference adaptive control. Proceed. 10th IFAC World Congress, Munich 1987, pp. 74-79.

[79] Safanov, M. et al.: Synthesis of positive real multivariable feedback. Int. J. of Control **45** (1987), pp. 817-842.

[80] Grimble, M.J.: Optimal H_∞ multivariable robust controllers on the relationship to LQG design problems. Int. Journal of Control **48/1** (1988), pp. 33-58.

[81] Haber, R., Keviczky, L. and H. Unbehauen: Applications of adaptive control on nonlinear dynamic processes - a survey on input-output approaches. Proceed. 10th IFAC World Congress, Munich 1987, pp. 1-14.

[82] Jedner, U. and H. Unbehauen: Intelligent adaptive control for a class of time-varying systems. Proceed. 12th IMACS World Congress, Paris 1988, pp. 377-380.

[83] Unbehauen, H. and Ch. Schmid: Status and application of adaptive control systems. Aut. Control Theory and Application 3 (1975), pp. 1-12.

[84] Parks, P. et al.: Applications of adaptive control systems. In Unbehauen, H. (Ed.): Methods and applications in adaptive control. Springer-Verlag, Berlin 1980.

[85] Schmid, Ch.: Design packages for adaptive controllers. In Singh, M. (Ed.): Encyclopedia of sytems and control. Pegamon Press, Oxford 1985.

[86] Litz, L.: Regel- und Steuergeräte auf der INTERKAMA 83, Teil 1, Chemie-Ing.-Technik **56** (1984), pp. A144-152.

[87] Kofahl, R. and K. Peter: INTERKAMA 86: Adaptive Regler. Automat. Praxis **29** (1987), pp. 122-131.

[88] Kurz, H.: ACHEMA 88: Regler und Reglersysteme. Automat. Praxis **30** (1988), pp. 544-550.

Expert Systems and Their Applications: A Survey

I. Hofmann, U. Meyer

Abstract

Beginning with the explanation of some basic principles, there will be shown representative chosen expert systems working in industries. Afterwards, the results of a project at the University of Bremen, the development of different expert system versions and the problems related to diagnosis of malfunctions of equiped printed circuit boards, are discussed.

1. Introduction

Because of increasing minimizing and the related explosion of powerness of computers software-tools with a knowledge based architecture come into account, named Expert Systems.

Contrary to classical problem solving methods based on numerical processing of variable data sets using fixed algorithms, the realisation of an expert system uses heuristics of a human expert about a special well known domain.

Therefor the main working area for expert systems is limited for those problems, which can hardly solved with numerical methods without closed results.Main points are tasks of diagnosis and analysis, problems of advisory and configuration and planing and supervisory tasks.

The main part of this paper shows some well known and new commercial expert systems especially for diagnosis and new results of own development in the area of knowledge based fault diagnosis of printed circuit boards.

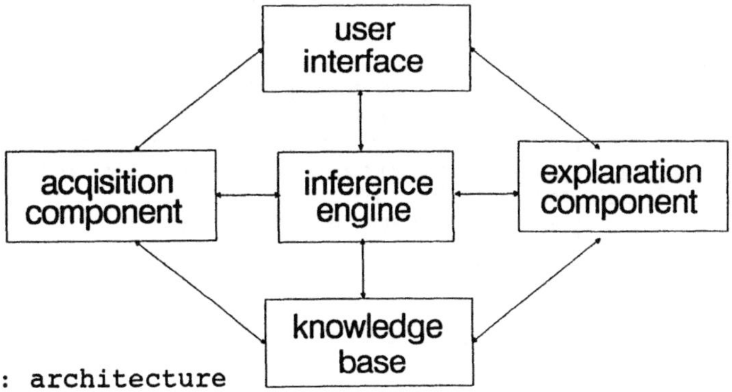

figure 1 : architecture

2. Architecture and characteristics of expert systems

Figure 1 shows the fundamental architecture of an expert system. Knowledge base and inference engine are the necessary components of an knowledge based system where the knowledge about the special domain is represented and where the inference mechanism for handling the knowledge is implemented. A submodul of the knowledge base, the dynamic database, is used as an actual memory during a consultation of the system.
The intersection between user and system is the user-interface, responsible for the dialogue.

The explanation- and the knowledge-acquisition-component are one of the main differences between expert systems an classical systems. The explanation-component should make clear the inferences for the user and the knowledge-acquisition-component has maintenance tasks for the knowledge base.

2.1. General demand to an expert system

The architecture of an expert system as shown in figure 1 indicates the main demand of a knowledge based system. In contrast to classical programming methods, where small changes of domain conditions normally require comprehensive changes of the programs, an expert system has to be built up in different moduls. By using modular methods, domain changes only need modifications of the knowledge base. Building up programs with empty frame-systems named shells, there should be the possibility to exchange the whole knowledge base without changing other system components.

The second demand concerns the explanation component of the system, which has the task to make the systems conclusions transparent and assessable for all users. The transparency is needed, because there is no user acceptance for expert systems which cannot explain their inferences and solutions.

	Semantic Networks	O-A-W-Triple	Frames	Production Rules	Modus Ponens	Horn Clauses
representation	objects, descriptors, relations	objects, properties	object frames with attribute lists and references	IF <condition(s)> THEN <conclusion(s)>	(A -> B) ^ A = B	BvA1[v..v An]
characteristics	closed problem description	part(s) of a semantic net concern. sp. object(s)	categorisation class hierarchies	inference, chain of reasoning, certainty factors	determination	disjunction with exactly one true literal

table 1 : knowledge representation forms

2.2. Structure of the knowledge base

Putting an expert system into reality, the main foundation is the selection of the right kind of knowledge representation. There are global differences between object-oriented, rule-oriented and logic-based knowledge representation. Table 1 gives a survey of the most important knowledge representation forms.

The characteristic of the object-oriented representation form is, that there exists a "Closed-World-Model" of the domain. Using frames for knowledge representation, that means to sort the attributes of the domain in different slots, is the most completest and most general method to represent complex domains.

The use of rules for representation has to be preferred if problem solution could be attained through generation of different states. Especially the use of production rules is qualified for planing problems.
For assessing the represented knowledge, importance factors or certainty factors can be used by programming with object- or rule-oriented methods.
These factors cannot be implemented without problems when you use a logic-oriented approach, because changing the heuristics into terms of predicate logic produces a determination of the possibly vague knowledge. At the other side a logic-oriented representation gives a strict consistent description of the problem. If it would be attainable to represent the domain with "horn-clauses" - that means disjunctiv coupled terms with one true statement - then a direct implementation with the language PROLOG is possible because that is the fundamental idea of PROLOG itself.

Handling complex problems with comprehensive knowledge bases often it is necessary to cut the knowledge into different parts. For high effectiveness it is possible to use different types of knowledge representation techniques for these different knowledge parts. As a rule such hybrid systems are in practical use.

3. Classification of expert systems

For classification of knowledge based systems in literature there exist a lot of partly different patterns. The classical attachment distinguishes only into four paradigms.

Diagnosis systems recognize applicable cases because of reduction of known data and give recipe proposals. One well-known older system of this pattern is the medical system MYCIN.

Advisory systems give dialog-oriented recommendations to handle the domain, most times to help a human expert. The advisory system PUFF is an example for this category (see chapter 5).

Configuration systems compose user-demanded structures through selection procedures. The best-known example for such systems is XCON (see chapter 5).

Planing systems are used for similar tasks, but beyond that they take time-dependent sequences into account. Examples are systems for generation of working schedules [3].
Later attempts to expand the classifications reach a more various differentiation of tasks of knowledge based systems, but also produce intersections of categories.

4. Applications of expert systems

Meanwhile industries and scientific disciplines, which use expert systems, are wide spread and cover a various diversity. At this point in particular only should be mentioned, that early systems mainly could be found in medicine and allied sciences like chemistry and biology or in computer science. In later times there exists a boom for developing advisory systems and planing systems for banking and insurance offices and for robotics.

Categorizing because of operational function areas, the area of application also varies. It is spread from buying via production and management up to marketing.

5. Some chosen commercial expert systems

Table 2 shows representative some well-known commercial used expert systems. Choosing these systems, the main regard was in the area of diagnosis systems.

6. Different methods for implementing diagnosis systems

Building a diagnosis expert system you can choose one of some different methods for implementation:

Starting with the domain-dependent knowledge you could choose the method of qualitative reasoning [Davis]. The basic idea is a hierarchical description of structures and functions. The system finds diagnosis because of deviations during comparison of behaviour between the test candidate and the model description of the system (deep reasoning). This kind of knowledge representation requires a high development expenditure.

Using <u>heuristics</u> (experience of a human expert) is another method to solve the problem. One possibility is the <u>systematic diagnosis</u> like traditional test systems are built. Often you only find an area where the solution has to be, but no real diagnosis. The processing time of such a system is high. The other possibility is the <u>symptom-oriented diagnosis</u> like the well-known system MYCIN was implemented. Based on heuristics and the in fact existing symptoms of the test candidate the diagnosis could be done goal-oriented and quick. In a restricted domain the development costs are limited.

Implementing a special diagnosis system at the Institute of Automation Technology we decided to use the symptom-oriented diagnosis to get acceptable results in justifiable time and because there existed a lot of knowledge and experience about the special test candidate.

SYSTEM	PROSPECTOR	XCON	PUFF	ACE	CATS1/DELTA	ES for LAN	MUD	YES/MVS
Institution	US Geolog. Survey	Digital	Stanford Univ.	Bell Labs.	General Electric	Lockheed	MIT	IBM
Objective	advice in mineral exploration	configuration of DEC comp. systems	decision support in diagnosis of pulmonary diseases	diagnosis of telephone nets	diagnosis and advice in repair of diesel-el.-locom.	diagnosis of a base-band subsystem	diagnosis and advice in oil exploration	real-time XPS for monitoring of MVS systems
Task	advice	configuration	advice, diagnosis	diagnosis, advice	diagnosis, advice	diagnosis	diagnosis, advice	diagnosis
Architecture	dialogue-oriented comparison analysis	configuration by client-specific requirements	generation of a diagnosis by probandum measurements	error detection + localisation by comparison	generation of malfunction hypothesis and repair proposals	evaluation of BITE-diagnosis notifications	diagnosis of drill liquid conditions	undertake of routine tasks in the MVS system
Tool	KAS (Knowledge Acquisit. System)	OPS5	primarily LISP, re-implemented in BASIC	OPS4	fig-FORTH	LES (Lockheed Expert System)	OPS5	modified OPS5
Reference (Year)	1984 (Prototype 1976)	1982 (Prototype R1, 1980)	1981	1983	1984	1984	1984	1984

table 2 : commercial used expert systems

7. Development of an expert system at the institute

The Institute of Automation Technology developed an expert system for diagnosis of malfunctions of equiped printed circuit boards. The system should support the finding of causes of malfunction during repairing of defective printed circuit boards.

First the knowledge base was implemented for one special printed circuit board - a special expensive A/D-converter, which belongs to a family of boards with similar structures. The system should be implemented so general, that other circuit boards could be diagnosed only by exchanging the description of problem-dependent knowledge, the knowledge base.

The diagnosis should be goal-driven and symptom-oriented. The human expert describes his experiences. He associates possible errors of the test candidate with its symptoms. He tells about suppositions of malfunctions, about exclusions of defects and about probabilities of frequency of malfunctions.

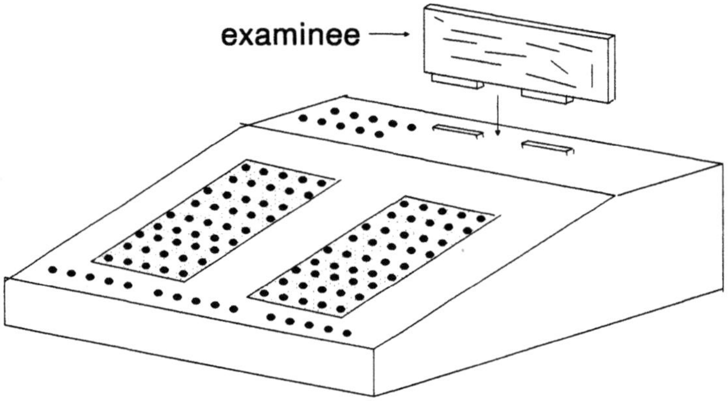

figure 2 : function tester

Stimulation and testing of the printed circuit board is done with an external function tester (figure 2). This function tester is coupled with the test candidate only by bus- and power plugs. To be universal, a testbed or a testpin is not in use. The system uses different test sequences to stimulate the circuit board and to get special symptoms of behaviour. These logical results - e.g. something would be out of tolerance or signals would be missed - is the input of the AI-computer (expert system). There exist two different solutions: the <u>offline-solution</u>, where the human user acts as an interface between function tester and expert system, and the <u>online-solution</u>, a direct coupling of both computers.

The formalisation and implementation of the knowledge was done in some different ways:

When applying the first two versions, the implemented knowledge of the system was based on natural language rules of an expert, concerning the test possibilities of a special circuit board by measurement. The rules are implemented in a search tree, using the language PROLOG.

Due to the inflexibility of these two versions, the knowledge base was re-implemented and the main result was a new system structure as basis for future works.

8. Basic system structure

Core of the system is at one hand the problem-oriented knowledge base and at the other hand the dynamic data base. In the knowledge base there are stored all descriptions about symptoms and diagnosis of the human expert, the test sequences of the function tester and repairing instructions. The data base gathers all special data of one diagnosis given with the help of the data acquisition. The inner core between the two bases is the inference engine. It finds conclusions through comparison of malfunction descriptions of the knowledge base and existing symptoms gathered in the data base. As long as there is no safe diagnosis, the inference engine establishes hypothesises about the cause of malfunction (see figure 3).

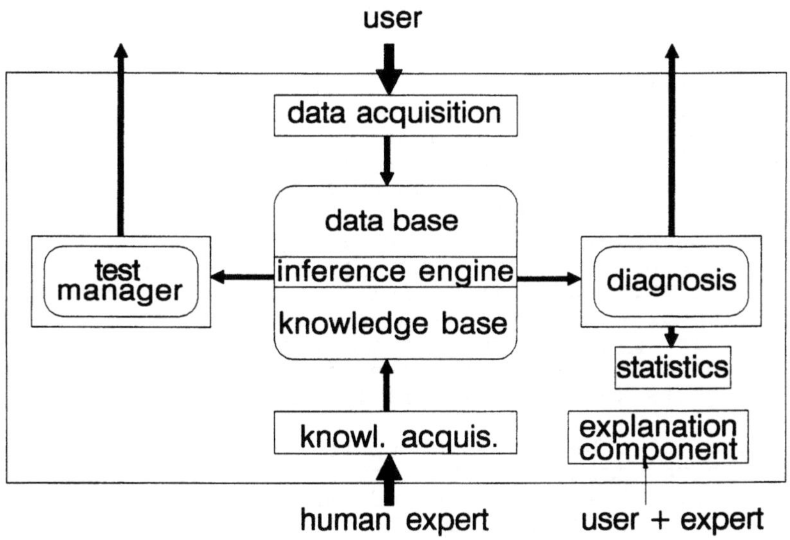

figure 3 : basic system structure

The test manager interprets the most important hypothesis and chooses the next test sequence to prove the hypothesis.

If a diagnosis has been established, the user gets the results together with repairing instructions. Afterwards statistics will be updated for quality saving and to improve new experience.

An acquisition component for modification and building of knowledge bases and an explanation component for clearness are implemented too.

9. Further developments of the system

Based onto the structure mentioned before a third version of the expert system was implemented. The domain knowledge was represented in the form of symptom-diagnosis-tupel as facts. The inference machine is working with pattern matching. The facts of the current examinee and the tupels of the knowledge base are compared. A conflict-set of all possible diagnoses is set up and the conflict resolution takes place through "hypothesize-and-test":

- For each element of the conflict-set an importance factor is calculated. The importance factor consists of a certainty factor and future diagnosis costs to diagnose an element as faulty.
- The element with the highest importance factor is imposed as hypothesis.
- The chosen hypothesis is verified by calling a new test sequence.

Therefor, the inference component is complex but possesses a comfortable and quick strategy for searching. Because of the implemented structure, the explanation- and the acquisition-component could be built up comfortable too.
In order to get the ability of comparison, a fourth version was implemented, where the only form of represented knowledge were production rules. The inference engine works like a normal production system. The main advantages of that implementation were little expenditure for the inference component and that no longer whole symptom-diagnosis-tupel have to be named. The main disadvantages of that implementation form were the loss of a fine strategy, the difficult explanation of diagnoses and there was no consideration of future costs.

The conclusion of the research up to now is to use the quickness and high comfort of the fact orientation at one hand and to use the flexible diagnosis and easy knowledge representation of the rule orientation at the other hand. A combination of both forms has to be done.

10. The final version

The foundation of the <u>fifth version</u> of the expert system for diagnosis of malfunctions of equipped printed circuit boards is heuristic knowledge about diagnosis with the function tester. To expand the capabilities of the system, an <u>extension</u> is implemented with other knowledge classes. The whole system is built up using a <u>blackboard architecture</u>. The data of current interest (diagnosis facts, hypothesis, ...) is gathered in the dynamic data base for communication with the inference engine. The knowledge base is filled with all data about diagnosis similar to the versions before and in addition there is stored knowledge about <u>functions</u> of PCB-parts, e.g. about C-MOS IC's and how to test them. Used data is wiped out and will be stored in <u>memory</u>.

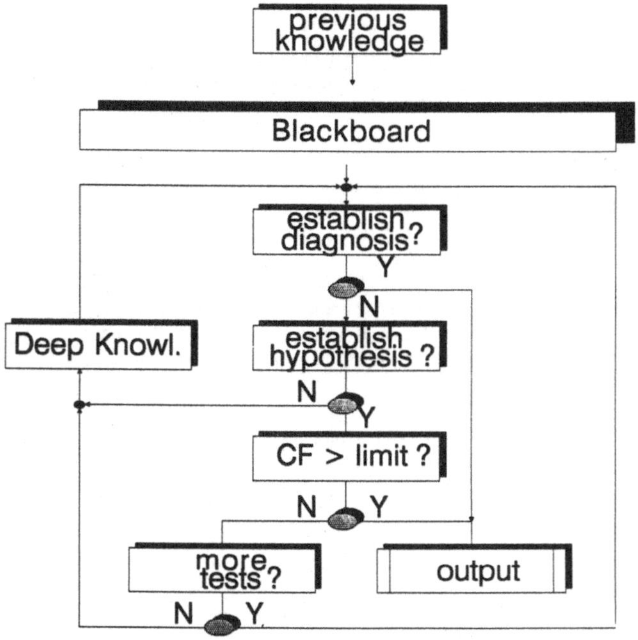

figure 4 : flow of diagnosis

The flow of diagnosis (figure 4) is based on blackboard data. First it is tested if there could be established a diagnosis and if this becomes true the diagnosis and repairing instructions would be offered by the system.

If there would be no diagnosis at the moment the inference engine tries to find a hypothesis about the cause of malfunction. If the importance factor of a hypothesis is greater than a limit, the hypothesis is offered.

If there exists a hypothesis with a small importance factor, normally the next necessary test sequence of the function tester is chosen and would be done to get new data about the examinee and to start a new inference sequence.

If there are no more possible tests of the function tester to find the malfunction of the examinee, the deep knowledge about functions of PCB-parts is used to get new data and to solve the problem.

If there exist no hypotheses, special knowledge about how diagnose the malfunction of the PCB is used to get new data.

11. Conclusions

Nowadays expert systems are spread above a big field of applications. The diagnosis is one of the possibilities to get advantages by using expert systems.

The research showed, that PROLOG is a powerful tool to program flexible expert systems for diagnosis. For a quick diagnosis of PCB's heuristic knowledge is favourable. If heuristics are not enough, switching over to function testing of single components is useful.

Tests showed, that the developed system is powerful, but it may be convenient to expand it with some other strategies, for example to include qualitative reasoning features for high performance.

The essential conclusion of the project is that using expert system techniques for diagnosis appears to be much more profitable than traditional techniques of systematic testing.

Literature

[1] A.Barr, E.A.Feigenbaum (Ed.); The Handbook of Artificial Intelligence, Vol. 1; W. Kaufmann Inc.; Los Altos; 1981.

[2] M.P.Georgeff; Strategies in Heuristic Search; in: Artificial Intelligence, An International Journal; Vol. 20; Elsevier Science Publishers B. V.; North Holland

[3] P.Mertens, K.Allgeyer, H.Däs, M.Schumann; Betriebliche Expertensysteme in deutschsprachigen Ländern; Universität Erlangen Nürnberg; Band 19, Nr. 6; Erlangen; 1986.

[4] I.Hofmann,I.Lorenz,C.Massau,A.Zimmermann; Forschungsbericht Nr. 1, Expertensysteme für Diagnosezwecke im technischen Bereich; Universität Bremen; 1986.

[5] I.Hofmann, A.Zimmermann; Forschungbericht Nr. 2, Expertensysteme für Diagnosezwecke im technischen Bereich; Universität Bremen; 1987.

[6] I.Hofmann, M.Petersmeier; Forschungsbericht Nr. 3, Expertensysteme für Diagnosezwecke im technischen Bereich; Universität Bremen; 1988.

[7] I. Hofmann; Systembeschreibung und Dokumentation der fünften PROLOG-Version des Expertensystems zur Leiterplattenprüfung; Universität Bremen; 1988.

[8] D. Popovic; Dokumentation eines regelbasierten Expertensystems zur Diagnose fehlerhafter, bestückter Leiterplatten; Universität Bremen; 1987.

Fibre Optic Communication Systems in Industrial Automation

D. Popović

1. Introduction

The fibre optic technology has since many years paved its way in most different field of activity of engineers. The original intention to apply it for noise-free data transfer has proven to be more promissing and of a good future so that it is to be expected that the future communication systems in the industrial and public sector will primarily be the optical ones. This will be shown here based on the main development streams: Local Area Networks, PROWAY, MAP, Field Bus, FDDI, Metropolitan Area Networks, etc.

2. Optical-Filter Communication

The first serious attempt to use the light-range frequencies for building communication systems has been registrated in early 60s as ruby crystal-based cohererent light generator was put into operation. The sharp light beam, generated in this way, could penetrate very deep into the surrounding and was thus applicable as an information carrier. At that time a series of other light sources, applicable to the same purpose, have been reported like semiconductor junctions, gas lasers etc. The point was to select the proper transfer medium to be optimally used for building of optical communication systems.

It was very soon identified that the glass could be the best candidate for this purposes, provided the losses (attenuation factor) in it could essentially be reduced. However, this was not an easy task to achieve. For instance, the first commercially available graded index fiber, produced in the late 60s, was able to guarantee a 100 dB/km loss, whereas the theoretical expectation have been fixed in the range of 20 dB/km. This was soon also achieved by the use of high silica glasses to produce the single-mode fiber. Nowadays, the fiber produced in this way can even go down with the losses up to 4 dB/km. Thus fiber-optic communication medium has grown up not only as a competitor to the conventional, copperbased medium, but in the mean time it has already overtaken it in the sense that the fiber optic medium has additional advantages like

- small diameter of the transmission cable
- light weight
- noise immunity
- wide band-with of communication channel
- low loss in signal transmission
- no crosstalk between adjacent signal paths
- no radiation and thus no interference or information tapping

However, the advantages have to be paid by coupling dificulties within the communication system.

The importancy of optical fiber for modern communication systems has had a decisive impact on development efforts in the material and design technology of silica fibers. Here, efforts are especially remarkable in the field of infrared optical transmisson media where heavy oxydes, fluorides, and chalogcnides are used a doping material to the glas fiber, lowering in this way the transportation losses down to 5 dB/km. Today, the engineers are working on realization of infrared fiber having 1 dB/100 km losses. Even with the plastic optical fibers one can implementet a communication link having around 20 dB/km losses.

Another important issue with this in connection is the optimization of fiber transmission performances, this because the silica fibers with the step index usually have the minimum losses at about 1550 nm window, whereas they have zero dispersion at about 1300 nm window.

Optimization means here to implement minimum los, minimum dispersion fiber with the same window. For instance it was shown that the fiber with a triangular klad will experience the minimum dispersion in the vicinity of 1550 nm window.

Fortunately, parallel to the progress in fiber optic technology just described, the advances in fiber production have been reportet so that we expect in the near future to have at the marketplace low-cost, high-performance infrared optical fiber.

Also in the field of intregrated optics the research works are going on succesfully. This should eliminate the present hybrid implementation of single mode and multimode optical fiber systems. Here, as a final result, the single semiconductor integrated circuit is expected in which electronic and optical devices will be integrated. The main development trends, registrated in this area, are opto-electronic, optical and monolitic integration.

3. Industrial Communication Networks

Modern factory automation technology is highly characterized by the extensive use of communication links for data and command exchange. This is expecially the case when applying multicomputer systems to build distributed process control or integrated plant control. Such links are in fact the core parts of the systems, thus being the main subject of research and development, which - because in the large-scale computer-based control systems a vast amount of information in various forms need to be transferred among the computers operators and some remote stations - is mainly concentrated on increase of data transfer rate (or of bandwith) of the comminication channels. Thus, also here, the problem solution was searched in possible application of fiber-optic communication paths for data transfer.

The most significant development trend, established in the early 70s, was directed towards improvement of relatively new, bus-oriented communication technology to replace the old, inefficient and costly point-to-point communication concept.

However, the application of a common communication link, be it a bus or a ring, supposes a high degree of standardization of equipment attached to it. This automatically implies that also the common communication link itself has to be standardized, including the protocols to be used for data transfer.

The standardization work for bus- and ring-oriented communication systems began really with the working out of a general model of such systems, known as OSI (Open Systems Interconnection) Reference Model of ISO (International Organization for Standardization) in the late 70s and early 80s. As it is well-known, the Reference Model is based on the seven layer koncept, whereby the layers are: physical, data link, network, transport, session, presentation, and application layer.

Based on this model, the standardization of specific communication systems, generally known as Local Area Networks (LANs), was able to start. It is to be reminded, that in late 70, the only commercially available and technically widely accepted communication system has been the ETHERNET of Digital Equipment Corporation, Intel, and XEROX. The system was aimed at interconnection of a wide range of equipment of different vendors by the use of a relatively inexpensive interface within a complex (or distributed) computer system. This, in fact, was the first, internationally non-standardized, local area network, based in CSMA/CD medium acces control technique. Up to 1024 devices can be attached to the system which data transfer rate is up

to ~10 Mbps. However, a rather low reliability of ETHERNET and the non predictable minimal access time of the bus have mainly been the obstacles for its application within the distributed computer process control structures. The first step in international standardization of a bus system to serve as a local area network has been done by IEEE in establishing the Project 802 (Local Area Network Standards) in 1980 which aimed at developing standard proposals for the two lowest layers of our Reference Model of ISO (e.g. for physical and data link layer).
Originally, the standardization work within the Project 802 was devoted to the baseband, bus oriented communication system with CSMA/CD medium access control procedure, like the ETHERNET was. Later extersion to token-passing and other procedures and system configurations has lead to the following partial standardization activities within the Project:

P 802.1: Overview and Architectures

P 802.2: Logic Line Control

P 802.3: CSMA/CD Access Method

P 802.4: Token-Passing Bus Access Method

P 802.5: Token-Passing Ring Access Method

P 802.6: Metropolitan Network Access Method

On the other hand, the IEC (International Electrotechnical Commission) has since 1975 been working on PROWAY (Process Data Highway), a communication link espicially suitable for data transfer within a plant automation system. The bus oriented system is a very reliable one, noise immune, of total length of up to 3 km. To the bus up to 100 participants can be attached to communicate with a rate of up to 1 Mbps by the use of HDLC (High Level Data Link Control) for formating of transfer messages. The final definition of PROWAY standard was published by IEC as Proway C in January (Publication 955). Along the same line, the initiative was made by General Motors in establishing the so called MAP Task Force, with the final goal to work-out a data transfer system suitable for interconnection of waht was known as "automation islands" of the production plant. For GM, this has concerned the integration of approximatelly 20.000 programmable controllers and about 2000 robots with the communication system to be standardized.
However, this standardization was not understood as a creation of new transfer or medium access procedures, but exclusively as a selection out of existing international standards those ones, that in the best way can be used within the MAP (Manufacturing Automation Protocol).
In the mean time, the MAP-User-Groups, pertaining over 200 internationally well-known MAP users and instrumentation suppliers, has issued the MAP-Ver-

sions 1.0 (April 1984), 2.0 (Febr. 1985), 2.1 (March 1985) 2.2 (August 1986) and Version 3.0 (April 1987). For widest industrial application, also a simplified version of MAP was standardized, the "MINI-MAP"-Version, which architecture includes the protocol layers 1,2 and 7 (Application Layer).
The evolution of data transfer technology in the past decade, supported by the ISO standardization of Local Area Networks, has implied that the selected MAP standards originate from ISO. The collection of such international standards representing MAP 3.0-Version are given in Fig. 1, along with the corresponding TOP 3.0 standards, the last standard concerning the Technical and Office (automation) Protocol, defining the standardized office communication system. Fig. 2 shows the internetworking (via the bridges and/or routers) of MAP and TOP Backbone Networks to produce a complex network, as needed for Computer Integrated Manufacturing (CIM).
The work on MAP international standards, a series of vendors of distributed computer control systems for plant automation has already integrated into their internal systems fiber- optic communication links or optical data

ISO - OSI	MAP 3.0	TOP 3.0	
Application	ISO DIS 8571 : FTAM (File Transfer, Access and Management) Subset ISO DIS 8649/1-2 and 8650/1-2 : ACSE (Application Control Service Elements) ISO DIS 9495 : Directory Services ISO DIS 9596 : Network Management ISO DIS 9506/EIA RS 511 : MMS (Manufacturing Message Service)	CCITT X.400 : MHS (Message Handling System) ISO DIS 9041 : Virtual Terminal ISO DIS 8613 : Document Interchange ISO DIS 8632 : Graphic Interchange	
Presentation	ISO IS 8822/8823 : Presentation Kernel ISO IS 8824 : ASN.1 (Abstract Syntax Notation One)		
Session	ISO IS 8326/8327 : Session Kernel, Full Duplex		
Transport	ISO IS 8072/8073 : Transport, Class 4		
Network	ISO DIS 8348/8473 : Connectionless Internet	ES-IS : ISO DIS	PLP : X.25
Data Link	ISO IS 8802/2(IEEE 802.2) : LLC1		HDLC LAP B (X.25)
Physical	ISO IS 8802/4 : Token Bus (5 MB Carrier Band) (10 MB Broad Band)	ISO IS 8802/3 : CSMA/CD (10 MB) ISO IS 8802/5 : Token Ring (4 MB)	X.21 / X.21bis

Fig. 1: MAP 3.0/TOP 3.0 International Standards

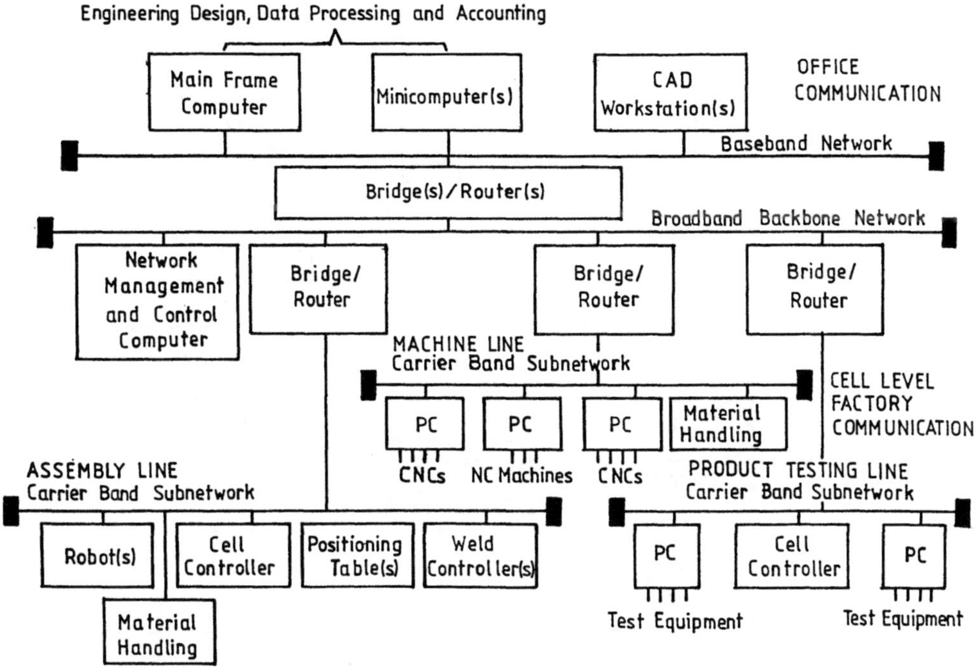

Fig. 2: Internetworking

highways. Fig. 3 shows the DCI 5000-System of Fisher and Porter in which the ETHERNET with the maximal length of 50 km is used as a basic communication link, to which the Direct Control Units connector. For covering longer distance, a Bus Repeater or a Remote Bus Repeater, operating in optical fibre as transmission medium, is used for point-topoint interconnection of two ETHERNET segments.

As a next example the MAX 1-System of Leeds & Northrup can be taken. It is a dual data highway system, based on a dual optical highway loop, to which the electrooptical interfaces are connected (Fig.4).

4. Field Bus

Recently, the field bus was proposed as a communication link between the process control computer and modern, intelligent, high-accuracy final elements (actuators and sensors), which in the mean time are provided with the capacity of remote calibration and telediagnostics. As its name shows, the bus is situated in the plant instrumentation field, so that for its implementation the optical fiber is prefered as data transfer medium. This, of

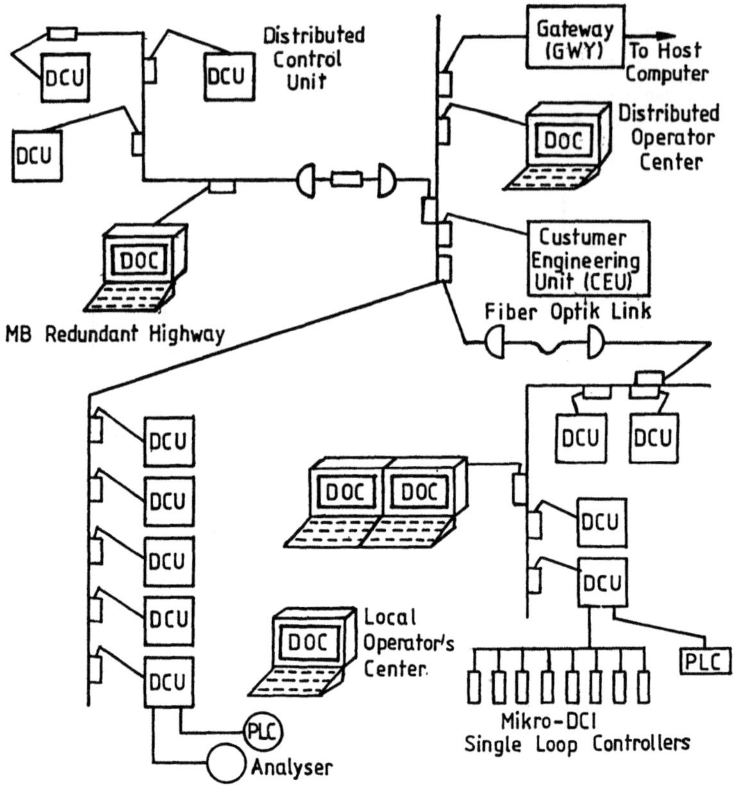

Fig. 3: Fibre-Optic-Based Multicomputer System

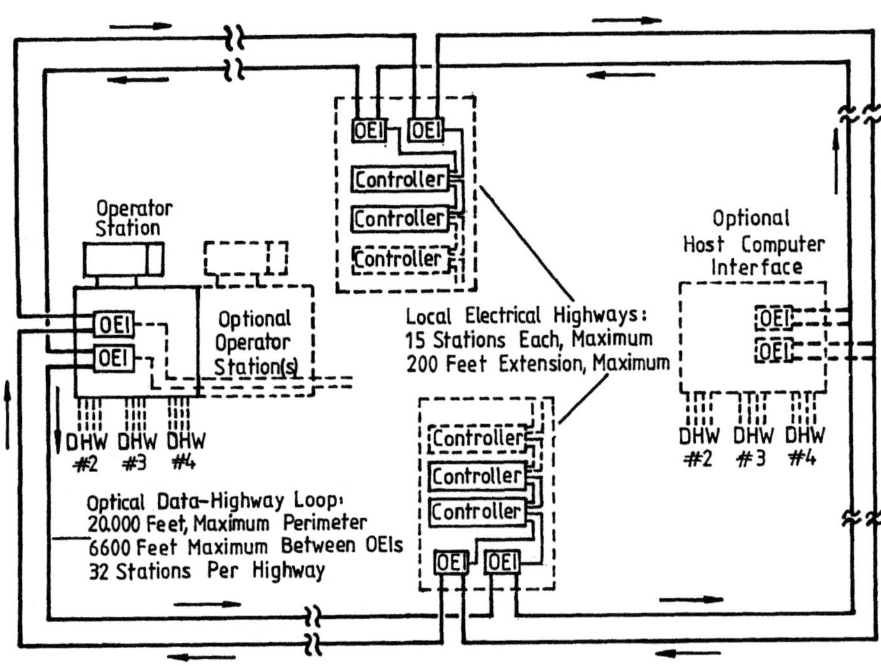

Fig. 4: Dual Optical Highway-Based Multicomputer System

course, mainly due to its noise immunity and explosion proofeness, but there are also other advantage of the bus, as compared to the classical, point-to-point final elements interconnection with the computer, like

- higher information content of the signals to be transferred, via the bus, including the calibration and diagnostic data
- lower cabling expences
- simplified interface of the computer to which the bus is attached

However, transfer of a large number of plant and control data via the common bus also introduces some technically unpleasant features into the system like

- lower reliability and availability of data transfer if no twin-bus-structure is used
- long response time due to the bus sharing
- complex electronics circuits to which the sensors and actuators are connected
- difficulties in powering of final elements, attached to the bus

Due to the special requirements and some specific restrictions within the plant, the field bus has to meet at least the following requirements

- independancy of bus participants from each other
- galvanic isolation of final elements
- plug compatibility with the standardized process instrumentation
- high reliability and high availability of the data transfer system
- powering of final elements

Most of the above requirements are in favor of the fiber-optic field bus. System redundancy can be increased by the use of a twofold bus (twinbus). In addition, cyclic redundancy code can be used for automatic check of received messages. Here, various efficient techniques are in use.
At the time being, international standardization work is going on within the IEEE (P 1118), IEC (TC 65), and ISA (SP 50) on specification of field bus parameters. Meanwile, the basic requirements of the future field bus draft standard have been elaborated, shown in Fig. 5.
From the table it follows that the field bus to be standardized should

- be a multidrop transfer system
- use both twisted pair, coax cable and optical fibre as transmission medium
- be 1.5 to 5 km long
- have not less than 32 participants

Standards Requirements	Standardizing Body		
	IEC TC65C	IEEE P1118	ISA SP50
Bus Topology			
Single Drop	—	—	Yes
Multiple Drop	Yes	Yes	Optional
Redundant	Yes	Yes	Yes
Transmission Medium			
Twisted Pair	Yes	Yes	Yes
Coaxial Cable	Yes	Yes	—
Optical Fibre	Yes	Yes	Yes
Bus Length	<1.5 Km	2 - 5 Km	<1.8 Km
Bus Participants			
Number	>30	<255	32
On-line Connectivity	Yes	Yes	Yes
Bus Arbritration			
Single Master	Yes	Yes	Yes
Multiple Master	Optional	—	—
Transmission Technique			
Broadcast	Yes	Yes	Yes
Multicast	—	Optional	—
Access Time	5 - 20 ms	10 - 50 ms	—
Scan Rate	—	—	>10, >100 /sec
Messages Rate	150 - 10,000	—	—
Error Detection	Yes	Yes	Yes
Signals Isolation			
Electrical	>250 V	>250 V	>250 V
Galvanic	Yes	—	Yes
Final Elements Powering	Optional	Optional	Optional

Fig. 5: Field Bus Draft Standards

- support redundant data transfer and twin-bus configurations
- have electrical isolation higer than 250 V
- have the possibility of final elements powering

The data transfer itself must also possible in broadband mode, enabling the access of 20-50 m sec. Presently, there are four outstanding candidater for future international field bus standard:

- BITBUS of Intel
- FIP of French origine
- MIL-STD-1553 of US Air Force/Navy
- PROFIBUS of German origine

All candidates, with the exception of BITBUS, have broadcast transmission technique and the multimaster bus arbitration method with the hybrid medium access control procedure.

It is of interest to point out, that the bus communication protocol makes use only from physical , data link, and application layer of OSI Reference Model of ISO.

5. Fiber Distributed Data Interface (FDDI)

With the enforced requirements of simultation factory and office automation, the demand was increased to introduce a Local Area Network capable to interconnect all essential parts of a company, e.g. its production, quality control, administration, etc. This, however, requires the simultaneous transmission via the same communication links of data, voice, and pictures, e.g. it really requires a relatively wideband communication system which, in fact, can only be a fiber opticbased system.

From all alternatives in the area of standardized fiber-optic communication systems, the widest international arceptancy has found the FDDI (Fiber Distributed Data Interface), on which X3T9.5 committee of ANSI (American National Standards Institute) was working for years Fig. 6.

FDDI is a 100 Mbps fiber optic Local Area Network,(FOLAN) mainly consisting of two rings of optical fiber, capable to interconnect up to 500 nodes (station). The interconnections between the neighboring nodes are pointto--point conncections, along which the in the nodes enhanced communication signals are propagated. The two rings, the primary and the secondary ring,

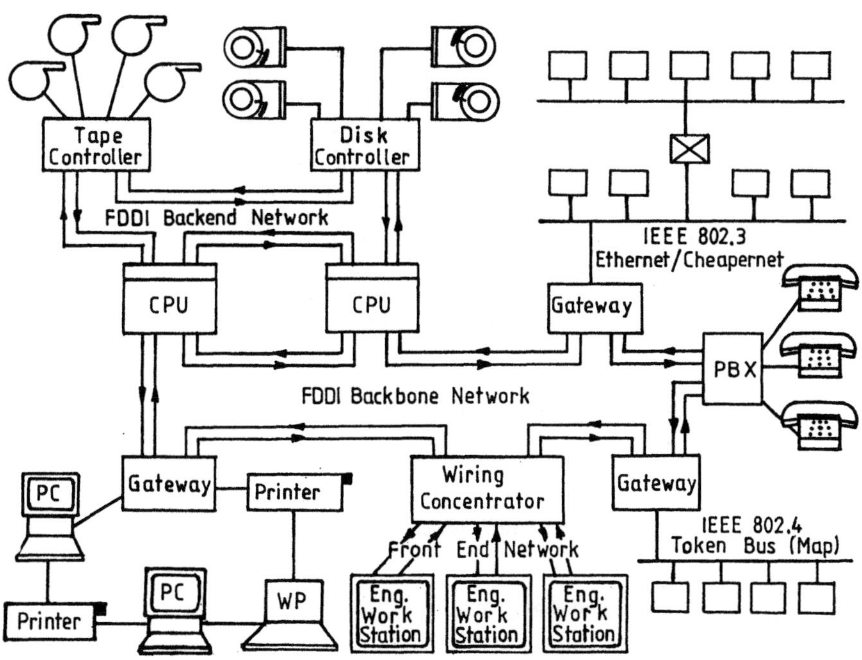

Fig. 6: FDDI Application

are used for data transmission, the last one being also able to serve as a back up ring in the event of link or station failure. FDDI can support two classes of stations:

- single-connected station, attached to only one ring,
- dual-connected station, connected with the both rings and, in the event of faults, reconfigurable

In general, recovery from ring failure is supported both by

- optical bypass switches, enabling the inactive node to pass the travelling light directly, e.g. without active power, from one to the other adjacent neighbors
- secondary ring as a redundant link which, in the event of link or node failure, can pass the data received back along the ring to the nearest logical station

In Fig. 7 the OSI-Reference Model of ISO of FDDI Standard is shwon, that directly concerns the two lowest layers only.

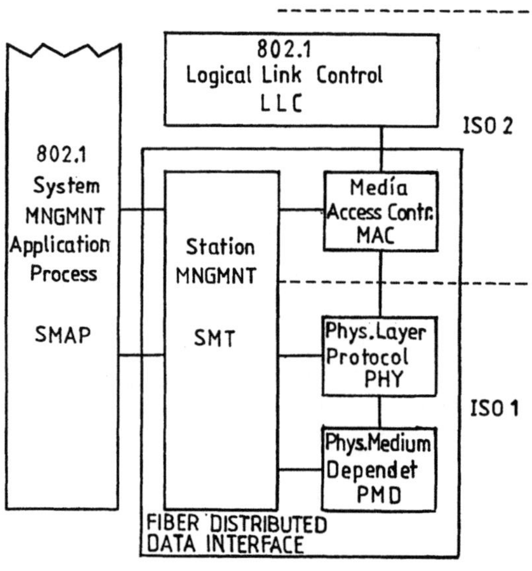

Fig. 7: OSI-Reference Model of ISO of FDDI

Basic FDDI characteristics can be summarized as follows:

- 100 Mbps transfer rate
- group encoding - 125 M Band
- sustained data rate 10 Mbps

- frames up to 4500 Bytes
- optical fiber medium
  - led-based technology
  - 2nd window transmission
  - duplex cables and connectors
  - 6 fibre sizes for cable plant
- Timed token rotated protocol
  - synchronous frames
  - asynchronous frames of multiple priority levels
  - restricted token mode
- dual ring connection
  - redundancy (dual Physical Layer)
  - dual access (dual MAC-protocol)
- optical bypass switch
- default timer values accommodate
  - up to 500 stations
  - up to 100 km cabled fiber
  - links up to 2 km

A recent enhanced version of FDDI the following capabilities:

- single-mode fiber media
  - links up to 60 km
  - potential for future higher data rate

Among the FDDI chips developed for commercial distributionor and available at the market, the best-knwon is the AMD's SUPERNET chip-set (Fig. 8) consisting of a

- RAM Buffer Controller
- Data Bath Controller
- Fiber Optic Ring Meara Access Controller
- Encoder/Decoder
- Data Separator

The chip-set is implemented in bipolar and CMOS technology, whereby the bipolar technology is used for high-speed physical interface and the CMOS technology for link layer physical and buffering functions.

The physical interface operates at 125 MHz and uses the low-power bipolar, low-noise switching technology. A flexible interface is used to fully implement the FDDI standard. At the time being, the SUPERNET can transfer data at the rate of 12.5 MBps, requiring for this purpose a bus bandwith of at least 25 MBps.

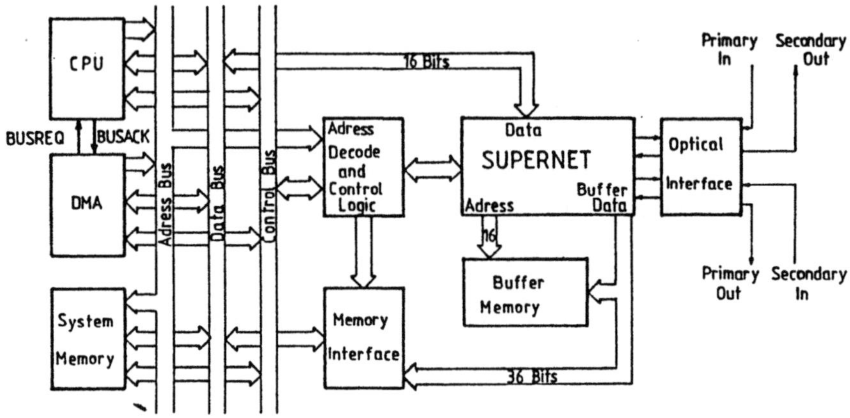

Fig. 8: SUPERNET FDDI-Set of AMD

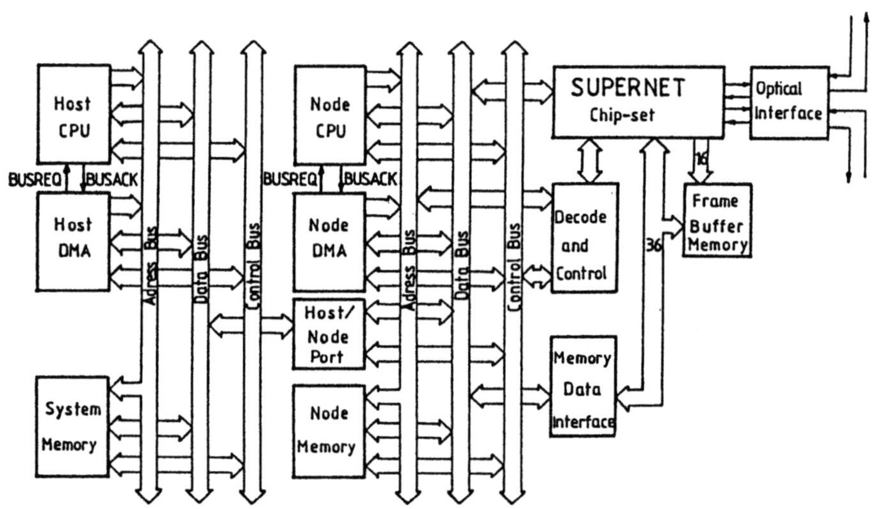

Fig. 9: Host-Processor-Based FDDI-Configuration

A typical low-cost system solution, relying on the host processor for protocol processing is shown in Fog. 9, whereby the possible configuration of a node controller is shown in Fig. 10.

A typical application exmaple of FDDI s shown in Fig. 11, where a building LAN, a data concentrator etc., situated in four different buildings, are interconnected by FDDI.

Although marketed in 1988/89, FDDI product will soon dominate at the data communication market. The prognostic for the period of time 1989/1992, Fig. 12, shows that the market potential of FDDI is enorm and rapidly growing.

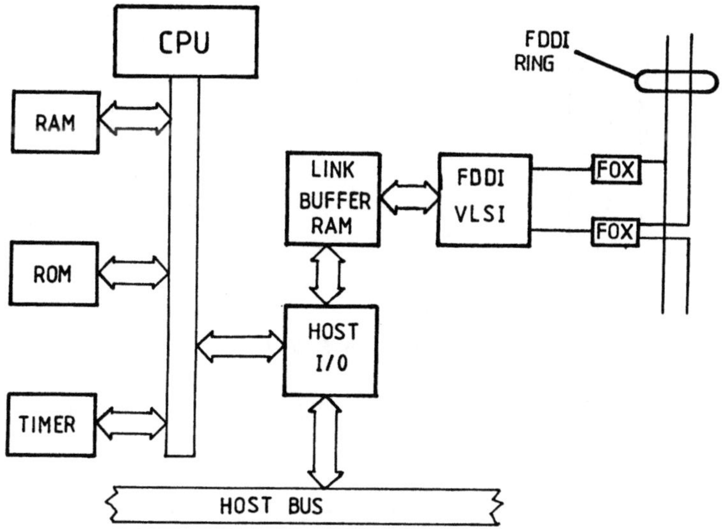

Fig. 10: Node Controller of FDDI

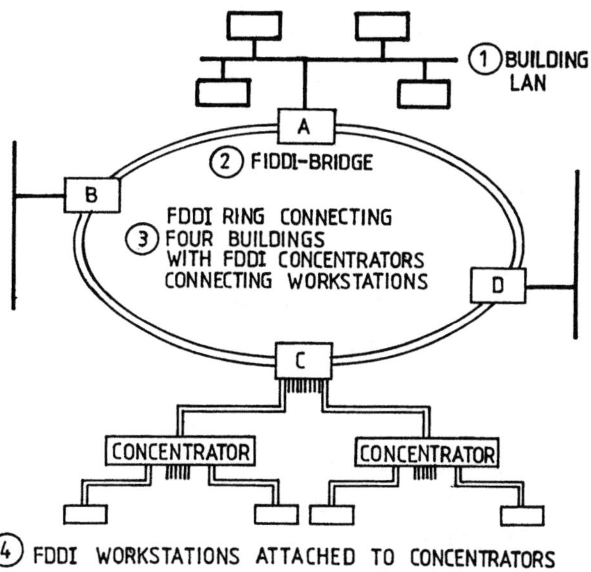

Fig. 11: Typical FDDI Application

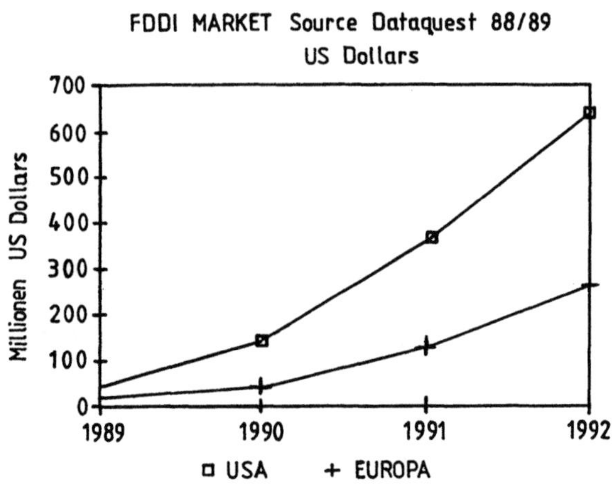

Fig. 12: The FDDI Market Prognostics for '89/'92

Presently, the work is going on standardization of an essentially improved version of Fiber Distributed Data Interface, FDDI II, which main characteristcs should be:

- circuit switched mode (added)
- dynamic bandwith assignment
- isochronous channels (16), each full-duplex, incrementally assignable and with the transfer rate of 6.144 Mbps
- suballocatable channels (for providing a variety of services)
- residual 0.75 Mbps data packet capability
- data steering logic in hardware

The time-priority of messages are

- isochronous data
- synchronous packets
- synchronous packets.

The application fields of FDDI-II include those of FDDI-I, and add the integrated LAN services like transfer of

- voice
- video
- real-time graphic display
- continuous sensor data systems
- control data streams
- distributed PABX networks

Fig. 13 shows the block-structure of FDDI-II standard.

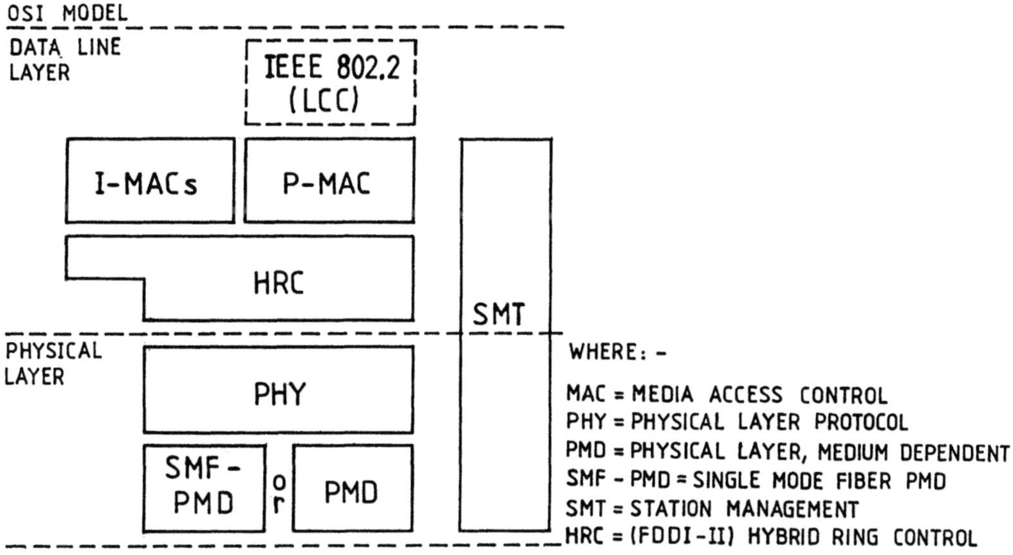

Fig. 13: Block-Structure of FDDI-Standard

Closely related to LANs and FDDI is the standardization and future use of Metropolitan Area Networks (MANs). A MAN is defined as a network that connects communication systems on non-contiguos properties within a distance range of a "metro area". It is a discrete, high-speed network for voice, data, and video, and provides LAN-to-LAN or LAN-to-WAN connections for public of private communication systems on non-contiguos properties within metropolitan distance range. The network definitions for a standard are being worked out within the IEEE 802.6 (Metropolitan Area Network) and will be approved by IEEE at the begin of 1990. As the most suitable candidates for implementation of future MANs are

- FDDI, most probably its version II, and
- DQDB (Distributed Queued Data Interface) sponsored and adapted by AT&T Network System

Similarly to FDDI, DQDB also uses two rings, Fig. 14, which is in fact a dual ring (logically even a dual bus), using the time slots technique for data on an "as-needed" basis. The interface is seen as a more suitable candidate than FDDI.

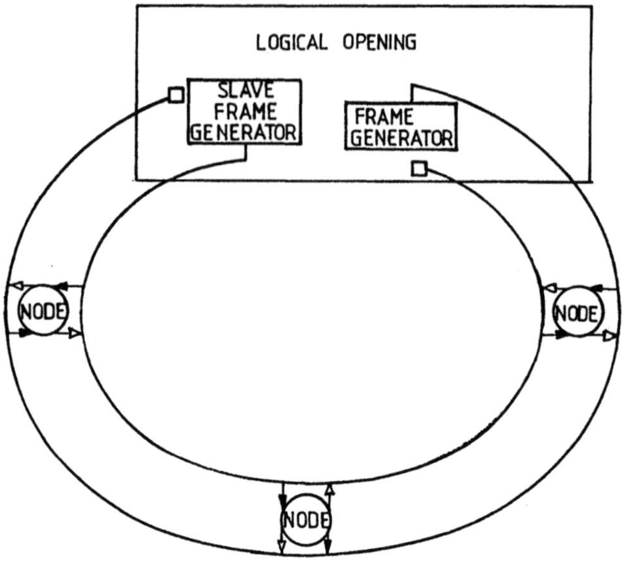

Fig. 14: The DQDB Architecture

6. Conclusion

It was shown that the optical fiber technology has the widest application within the industrial sector: from the field bus for measured data acquisition, local system busses and remote highways, up to complex, multimedial interconnections that also include simultaneous voice and video transmission. The distributed structure of a modern production company having a number of production, quality control, inventory, administration, and salesment buildings, including the within a metropolitan area spread out extensions, requires a flexible, multimedial communication system that in the time to come, will most probably be FDDI-II-based.

7. References

1. Ezekiel, S. and Araitty, H.J. (1982). Fibre-Optic Rotation Sensors, Springer-Verlag, New York.
2. Fibre Optics in Adverse Environments (1981). SPIE Vol. 296, Conf. at San Diego, 1981.
3. Optical Fibre Sensors, (1983). Intern. Conf. at London, IEEE, Vol. 221.

4. Woolvet, G.A. (1977) Transducers in Digital Systems, Peregrinus, U.K.
5. Day, J.D. and Zimmerman, H. (1983). the OSI Reference Model, Proc. IEEE, 71, pp. 1334-1440.
6. Graube, M. and Mulder, M.C. (1984). Local Area Networks, IEEE Computer, 17, No. 10, pp. 242-247.
7. Tolhurst, M.R., Ed. (1988). Open Systems Interconnections, Macmillan Education Ltd., London.
8. Agarwal, D.C. (1988). Fibre Optics in Communication, A.H. Wheeler & Co. Pvt. Ltd. New Delhi.
9. Baker, D.G. (1985). Fibre Optics Design and Application, Reston Publishing Co., Reston, Va.
10. Chevin, A.H. (1987). An Introduction to Optical Fibers, McGraw-Hill Internat. Eds., Singapore.
11. Keiser, G. (1986). Optical Fiber Communications, McGraw-Hill Internat. Student Ed., Singapore.
12. Kopany, N.S. (1987). Fiber Optics, Prinziples and Applications Acad. Press, New York.
13. Advanced Micro Devices. (1987). The SUPERNET Family for FDDI, Data Handbook, AMD, Sunnyvale, Ca.
14. Baker, D.G. (1986). Local Area Networks with Fibre-Optic Applications, Prentice-Hall, NJ.
15. ISO (1988). Fibre Distributed Data Interface (FDDI), Proposed Standards: Token Ring Physical Layer Protocol; Token Ring Media Access Control (NAC); Physical Layer Medium Dependent (PMD), ISO 9314-1,2, and 3.
16. Pleinevaux, P. and Decotignie, J.-D. (1988). Time Critical Communication Networks: Field Busses, IEEE Network, 2, No. 3, pp. 55-63.
17. Warrior, J. and Cobb, J. (1988). Structure and Flexibility for Fieldbus Messaging, Control Engineering, 35, No. 10, Vol. 2, pp. 18-20.
18. McGuffin, L.J., Reid, L.O. and Sparks, S.R. (1988). MAP/TOP in CIM Distributed Computing, IEEE Network, 2, No. 3, pp. 23-31.
19. Schutz, H.A. (1988). The Role of MAP in Factory Integration, IEEE Trans. Industr. Electronics, 35, No. 1, pp. 6-12.
20. Valovic, T. (1989). Metropolitan Area Networks: a Status Report, Telecommunications, 23, No. 7, pp. 25-32.

Route Optimization Algorithms for Expert Systems

Freimut Hillert

Abstract

For maximization of a weight function over a system of indepedent sets a new type of algorithm had been found:
It could be proved, that the choice of elements with "local" maximal weight can construct the same near-optimal solution as selection of global maxima. This reduction of saearch area enables a drastic decrease of CPU-time needed for solution. In this paper it is discussed, how to use such type of algorithms for solving one-depot or multiple-depot vehicle routing problems and how to integrate such algorithms in expert systems.

1. Introduction

Within the framework of developing an expert system for Route optimization and decision-making the following task had to be solved:

> One or more vehicles (cars, ships, airplanes, ...) have to visit particular places (i.e towns or citties) from one or more points or depots - each destination being served by just one vehicle - to carry out service tasks. For this, convenient routes have to be found and planned in detail, taking into account the influence of unexpected disturbance variables.

This resulted in a classification into three partial tasks:

- Routing (optimization of routes for minimum distance)
- Scheduling (detailed sequence planning based on the routes determined) and
- Controlling (control and regulation of navigation sequence considering the disturbance variables)

The following optimization problems had to be solved for the first of these tasks:

Travelling salesman problem (TSP):

n Points P_1 ... P_n have to be visited from one depot P_o by just one vehicle. A route has to be found which is as short as possible.

Should one vehicle be able to carry out only a limited number of the service tasks required and this capacity is not sufficient for all the jobs to be dealt with by the one vehicle, the result is the "classical"

Vehicle routing problem (single depot, multiple vehicle, node routing problem - VRP):

Points P_1 ... P_n have to be visited from one depot P_o to deal with service tasks ranging Q_1 ... Q_n. For this, suitable vehicles with capacity K are available in adequate numbers. Routes have to be found such that, along each route, capacity K is not exceeded by the sum of the service tasks Q_i and the shortest possible overall travelling distance is reqired passing through each point just once. Distances c_{ij} between Points P_i and P_j are taken to be known.

If the vehicles are distributed amongst several depots, we have the

Multiple depot vehicle routing problem (MVRP),

which differs from VRP in that k depots P_o^1 ... P_o^k have to be considered.

A resolving Algorithm has been developed and implemented for the latter problem which is also suitable for solving the other two problems mentioned as these are themselves special cases of that problem.

Where, in the following, an account is given of the results of investigations into this problem area, it should be assumed (in the interests of a more understandable representation) that this refers to VRP. Almost all of the results can be applied to MVRP without any appreciable alterations.

2. Savings methods for route optimization

In the "classical" VRP, there are two types of points with considerably differing quality: P_o, the depot, and all the other points to be visited from the depot. Whereas in the depot can start more than one used connection (one for each used vehicle) and also more than one used connection can lead to the depot, to each other served point leads just one used connection and also starts in such point just one used connection. As it is to be feared that this difference cold have a disturbing effect in the determining of a solution for the VRP, an attempt will be made in the following throught the elimination of point p_o as well as all of the connections starting from and leading to it to attain an equivalent problem with qualitatively equivalent points.

With this objektive, distances $c_{i,j}$ between points P_i and P_j are divided in such a way that a part of these distances is assignes to the starting point as well as to the destination point of the connection:

$$c_{i,j} = u_i + d_{i,j} + v_j$$

u_i and v_i are to be selected of all at will (but fixed) for the respective point P_i.

The following equation then applies for the total route length $l(L)$ of an allowable solution L to the VRP:

$$l(L) = \sum_{L} u_i + \sum_{L} d_{i,j} + \sum_{L} v_j$$

Where L contains m tours, (and n points $P_1 \dots P_n$ are to be visited,) the following applies

$$l(L) = \sum_{i=1}^{n} u_i + m \cdot u_o + \sum_{L} d_{i,j} + \sum_{i=1}^{n} v_j + m \cdot v_o$$

as, in L, precisely one connection leads to each P_i with $1 \leq i \leq n$ which is used in L, (and just one connection used leads away from P_i,) while this is the case m times for the depot P_o. Indepedence from the number of tours is achieved by establishing

$$u_o = 0 \qquad \text{and} \qquad v_o = 0.$$

$l(L)$ thus results from the sum of the "reduced" distances $d_{i,j}$ plus the (fixed) sum of the "potentials" u_k and v_k of all the points P_k with $k \neq 0$.

Elimination of the point P_o assigned to the depot becomes possible when the reduced distances $d_{o,i} = 0$ and $d_{i,o} = 0$ are assigned to all of the connections starting from or leading to P_o. This results in

$$0 = d_{i,o} = c_{i,o} - u_i - v_o \qquad \text{and}$$
$$0 = d_{o,i} = c_{o,i} - u_o - v_i$$

for all P_i with $i \neq 0$.

Because of $u_o = v_o = 0$:

$u_i = c_{i,o}$ and $v_i = c_{o,i}$ apply and thus

$$d_{i,j} = c_{i,j} - c_{i,o} - c_{o,j} .$$

The triangular inequality

$$c_{i,j} \leq c_{i,o} + c_{o,j}$$

means $d_{i,j} \leq 0.$

This therfore suggests the transition to the non-negative

$$s_{i,j} = -d_{i,j} = c_{i,o} + c_{o,j} - c_{i,j}$$

which are described as "Savings".

(This term goes back to CLARKE and WRIGHT who, in 1964, published a route optimization algorithm in which savings $s_{i,j}$ played a fundamental role, being established via the interpretation:

If Points P and P are visited from the depot via a tour P_o --> P_i --> P_j --> P_o instead of the targeted journeys P_o --> P_i --> P_o and P_o --> P_j --> P_o, routes P_i --> P_o and P_o --> P_j are dispensed with, whereas, on the other hand, route P_i --> P_j has to be travelled additionally. The saving calculated from this is $s_{i,j} = c_{i,o} + c_{o,j} - c_{i,j}$.)

VRP is thus equivalent to the task of finding tours which do not exceed capcity K and provide a greatest sum of savings. This concerns a special task of the maximation of a weight function over an system of indepedent sets. This task is one of the more difficult problems of the discrete otimization for which an exponential increase of the calculation time has to be reckoned with according to the size of the problem (the number n of points). When, as in the present case, a solution has to be established within a given period, approximation methods are implemented. As such approximation method the "simultaneous savings algorithm" has found widespread use for solving the VRP. This algorithm develops the solution step by step by attempting to add the connections between the points P_i and P_j to the solution in the descending order of their savings $s_{i,j}$:

(1) Find the connection P_i --> P_j with $i \neq 0$ and $j \neq 0$ which has the greatest savings value $s_{i,j}$ among all the connectionc not yet examinned.
Add this connection to the solution when
- this does not cause capacity K to be exceed,
- a circle wood not occur and
- after adding, not more than one connection still leads to and not more than one connection also starts from each point.

When all such connections have been examined:

(2) Connect all points with no connection starting from (or leading to) them to the depot P_o.

3. Local savings algorithm

As the solution of the VRP is to be effected within an expert system, there is a necessity for an explantory component. The question as to why a certain connection has been selected (and not another particular one) in the algorithm cited above, cannot be anwered with any great degree of satisfaction. The only available information is that connections with a greater savings value are preferred. Whether this means that another connection with a smaller savings value is excluded from inclusion in the solution first remains unclear. This is a reason for looking into the question as to which connections are excluded (or could be excluded) from simultaneously being a part of the solution.

In the following, the case will be examined of the inclusion of connection P_i --> P_j in the solution. A connection sequence P_k --> P_i as well as P_j --> P_l already exists in the solution in this regard. (No connection in the

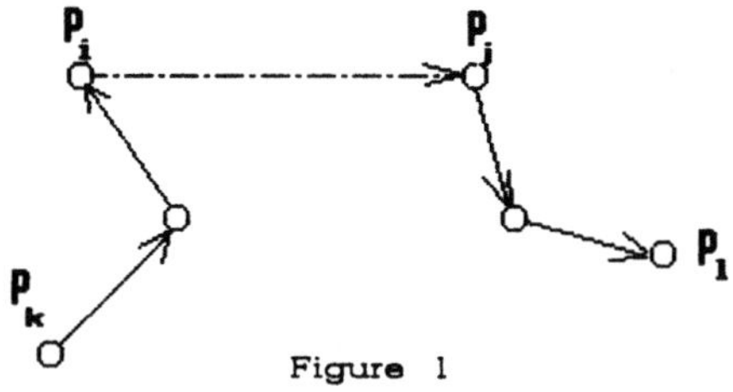

Figure 1

solution set should lead to P_k and P_j, just as no such connection should start from P_l or P_i - see Fig. 1.) Where the solution does not yet contain any such connection sequences, we would have k = i or j = l.

The connection P_i --> P_j can under no circumstances be included in the solution when

- this would cause load capacity K to be exceed or
- a circle would be formed (i.e. in the case of i = l or j = k).

Furthermore, simultanous inclusion of P_i --> P_j and P_i --> P_x ($x \neq j$) as well as of P_i --> P_j and P_x --> P_j ($x \neq i$) for random points P_x would, for structural reasons, be exceeded.

Finally, the inclusion of a connection P_x --> P_k or P_l --> P_x could exclude connection P_i --> P_j from the solution as this could possibly cause capacity K to be exceed.

The totality of all the connections which could (in the situation examined) mutually exclude themselves from inclusion in the solution thus consists, with regard to connection P_i --> P_j, of the connections P_x --> P_k, P_x --> P_j, P_i --> P_x and P_l --> P_x whereby P_x can be any point. This set of connections can, in certain way, be regarded as the neighbourhood of connection P_i --> P_j. Where, in this neighbourhood, saving $s_{i,j}$ has the highest value, connection P_i --> P_j is referred to as local maximal. It has been proved that the sequence in which local maximal connections are included in the solution is irrelevant, i.e. (where the savings each have different values) it does not influence the structure of the solution.

While, in the case of the simultaneous savings algorithm, a connection with a globally maximal saving must constantly be sought, an algorithm restricting this search to the local maximum, and referred to in the following as "local savings algorithm", diminishes the extent of the search area by approximately factor 4/n, thus being considerably more effective for the same result.

The procedure for a local savings algorithm can be described as follows:

(1.) Select any connection P_i --> P_j which cold be included in the solution as the starting point for the search. (Considerable computing time can possibly be saved through skilful selection of this starting point.)

(2.) If this connection P_i --> P_j is the local maximum (i.e. it has the greatest savings value in his neighbourhood), continue with **(4.)**.

(3.) If, in the neighbourhood of P_i --> P_j, there exists a connection P_x --> P_y with $s_{x,y} > s_{i,j}$ which could be included in the solution, select P_x --> P_y as new starting point (i := x ; j := y) and continue with **(2.)**.

(4.) Add connection P_i --> P_j to the solution. If there are still connections which could be included in the solution, continue with **(1.)**.

(5.) If there are no more connections which could be included in the solution, connect all the points with no connection starting from or leading to them to the depot P_o.

Proof of the functionability of such algorithm has also been furnished for the generalised problem (maximization of a weight function over a system of indepedent sets) so that it has also been possible to ensure applicability for multi-depot problems. As each global maximum is also a local maximum, the "classical" simultaneous savings algorithm (according to CLARKE and WRIGHT) represents a special variant of a local savings algorithm.

4. Savings algorithm, represented by rules

As has already been explained, the decision as to whether a connection is included in a solution L is equivalent to the proof of its local maximality. This local maximality requires, for its part, the fullfilling of two conditions:

(1) The connections must be "allowable", i.e. it may not go against the structural requirements of a solution and

(2) it should have the highest savings value among all of the allowable connections in its neighbourhood.

The following detailed rules can now be derived from these conditions:

(I) A connection is not allowable when its inclusion in the solution L would lead to the forming of a cycle (Fig. 2) or cause load capacity K to be exceed.

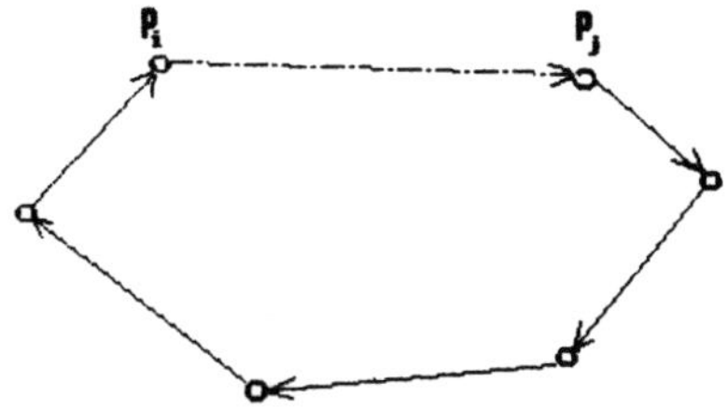

Figure 2

(II) A connection is not allowable when, would be its inclusion in the solution L, capacity K exceeded through th sum of the demands along a connection sequence in L.

(III) A connection is not allowable when it leads to a point to which a connection from L already leads . (Fig. 3)

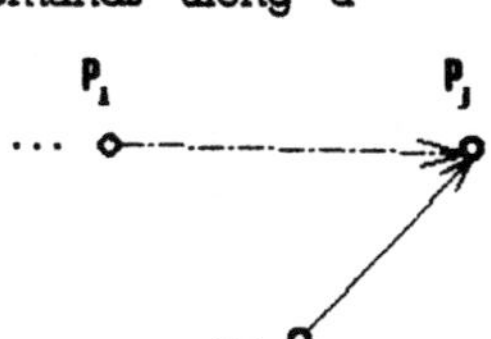

Figure 3

(IV)	A connection is not allowable when it starts from a point from which a connection from L already starts. (Fig. 4)

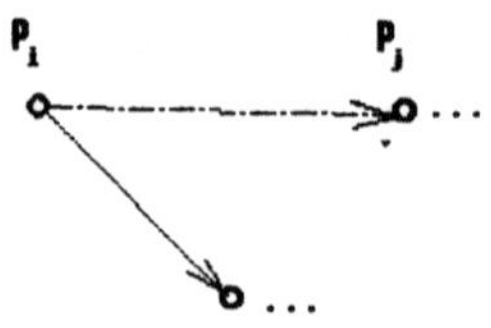

Figure 4

Only allowable connections can be included in the solution.

According to these four rules, which provide a complete description of the allowability of a connection (with regard to a partial solution set L already constructed), further rules can now be formulated describing which connection is to preferred to another. Two rules concern reliable connections which unreservedly exclude each other from inclusion in the solution set L:

(V)	If two allowable connections start from the same point, the connection with the lower savings value may not be included in the solution.

(VI)	If two allowable connections lead to the same point, the connection with the lower savings value may not be included in the solution.

A further rule concerns connections which could possibly (throught capacity K being exceeded) exclude each other from inclusion in the solution:

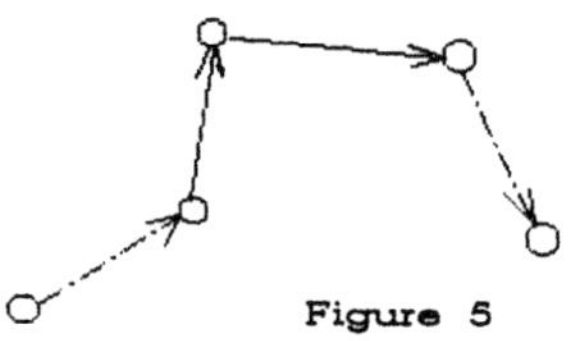

Figure 5

(VII)	If an allowable connection leads to the start of a connection sequence from L and a further allowable connection starts from the end of this connection sequence, the connection (of these two) with the lower savings value may not be included in the connection.	(Fig. 5)

These rules (I) to (VII) describe completely the local savings algorithm in relation to the reasoning for urgently necessary decisions and can thus form the basis of an explantory component. All of the others decisions to be made are not urgent and do not, as a rule, have any influence on the form of the constructed solution L.
If the coordinates of the points P_o ... P_n are available, the explantory component should, because of the greater clarity it acieves, be backed up by a graph on the terminal screen.
A solution of the VRP can, of course, be established with the help of these rules when they

are incoporated into an expert system. However, it is to be feared that, with this, the computing time required for the solution vis-a-vis realisation as the "classical" prozedural algorithm will increase.

5. Interactive optimization throught savings algorithm

Getting a detailed explanation, elucidation and reasoning of the algoritms sequence of decisions the user sometimes will recognise (actual or putative) "wrong" decisions. As the algorithm is an approximation method, it is in the nature of the things that not all decisions can be taken as optimal ones. If, when solving a determined optimization problem, one recognices at what point the decisions of the algorithm can be improved, the temptation will arise to add additional rules which prevent the respective wrong decision for the dertermined problem. If this procedure is continued on and on, the extent of the rules would increase infinitely without any possibility of an improvement of the algorithm in principle. The reason for it is - as proved by HAUSMANN, KORTE and JENKINS - that algorithms for the solution of problems of this type, for which the maximum deviation of the solution from the optimum turns out to be lower than for the savings algorithm presented, are associated with a level of caculation time, which increases exponentially with the size of the problem (the number n of points). The following table is designed to convey an impression of the speed of increase:

n	10	20	30	40	50	60
	$2.5 * 10^{-4}$ secs.	0.1 secs.	2.6 mins.	38.5 hours	4.028 years	3.766 thousand years

A method which would require such an effort to construct the solution of an VRP would no longer be usable, at least for more extensive examples. As, throught the continued adding of rules, it is therefore only possible to achive a gradual improvement of the method related to the respective example, a more rational procedure presents itself: the user is provided with the possibility of intervening in the algorithmic sequence at any point.
Such an interruption requirement can ensue throught the input of the numbers of two points, causing an interruption when a connection between these two points is to be included in the solution. At this point the user is given the possibility of acessing detailed information regarding the respective situation (using explantory component). Subsequentely he can change the structure of the solution by adding connections or departing them from the solution. In doing so, a safeguard must be ensured that none of essential structural requirements (see rules (I) to (IV)) can be violated. After the user has executed all the

planned changes, he can make the optimization algorithm continue the work - where applicable, up to the next planned alteration. If, instead of the numbers of the points, a "*" is also accepted as an input with the interpretation that the parity of "*" with any number is supposed, the interruption requirement can (throught the input of a nunber and a "*") also be applied to a point. Finally, an individual step sequence (attained throught the input of two asterisks) can be used effetively to demonstrate the algorithmic sequence.

6. Specifics in strategy for solving multi depot vehicle routing problems (MVRPs)

There are some different strategies for solving MVRPs recommended in scientific papers. TERNO (p. 73) interprets MVRP as optimization problem which is to solve in two steps. First of these steps decides an assignment of points $P_1 \ldots P_n$ to depots $P_o^1 \ldots P_o^k$. This task is interpreted and solved as "classical transport problem". (Additional can be considerated limited capacity of depots.) Optimal solution gives an assignment of points to depots (where some Points can be assigned to more than one depot and thus would be visited by more than one vehicle).

Routes are determined in second step separate for each depot by savings algorithm.

Such strategy requires small effort to construct the solution because dimishing dimension of problems which are to solve one after another in second step. On the other side this strategy is connected with important disadvantages. So such solution violates essential restriction of the MVRP visiting points by more than one vehicle and, furthermore, length of the tours grows longer.

GOLDEN et al. (p. 131) recommend assignment of points to depots in two steps: Points which have a great difference between distances to the closest and the second closest depot are assigned first to their nearest depot. The other points are than assigned to a depot in manner of least increasing total route length by inserting between two other points already assigned to the same depot.

BODIN et al. assign all points temporary to their nearest depot. Than they calculate savings

$$s_{i,j}^k = c_{i,o}^k + c_{o,j}^k - c_{i,j}$$

for each points P_i and P_j and each depot P_o^k and add to solution allowable connection with greatest savings value.

$c_{i,o}^k$ are calculated as

$$c_{i,o}^k = \min_t \{ \overline{P_o^t\ P_i} + \overline{P_i\ P_o^t} - \overline{P_i\ P_o^k} \}$$

if P_i so far is temporary assigned to a depot,

$$c_{i,o}^k = \overline{P_i\ P_o^k}$$

if P_i is permanent assigned to the depot P_o^k and

$$c_{i,o}^k = 0$$

if P_i is permanent assigned to another depot then $P_o^k \cdot c_{o,j}^k$ is calculated in the same manner. P_i and P_j are assigned permanent to depot P_o^k after adding P_i ---> P_j to solution owing to $s_{i,j}^k$ was the greatest saving.

Latter two strategies never revide permanent assignments of points to depots. This can, as will be shown by the following example, lead to bad results:

Fig. 6 shows a simple problem with four points and two depots.
Because of the greater savings values first connections P_1 ---> P_2 and P_3 ---> P_4 are added to the solution. They are assigned to the nearest depot P_o^1. (Fig. 7)
Assignment to depot P_o^1 is now permanent, it cannot be changed after adding next following connection P_2 ---> P_3. (Fig. 8)
Assignment to depot P_o^2 would be better in fact as is shown in fig. 9.

More rational algorithmic strategy, maked possible by low calculation time level of local savings algorithm, can be find by holding temporary all assignments to depots while whole algorithm. Best assignment to depot must be calculated once more after adding each connection to solution set. Difference in route length (before and after adding connection) is interpreted to be the saving and controls sequence of algorithm.

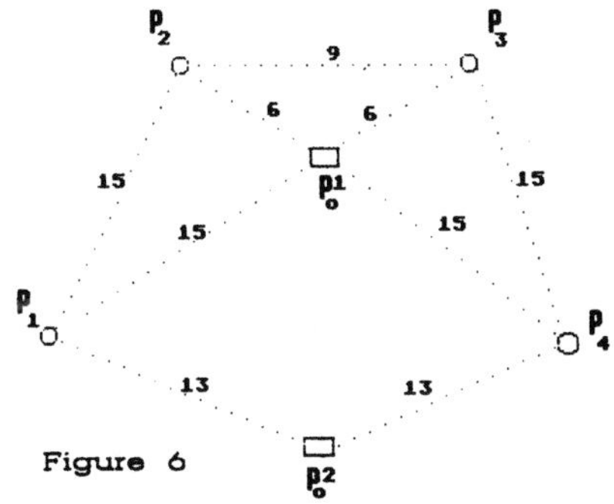

Figure 6

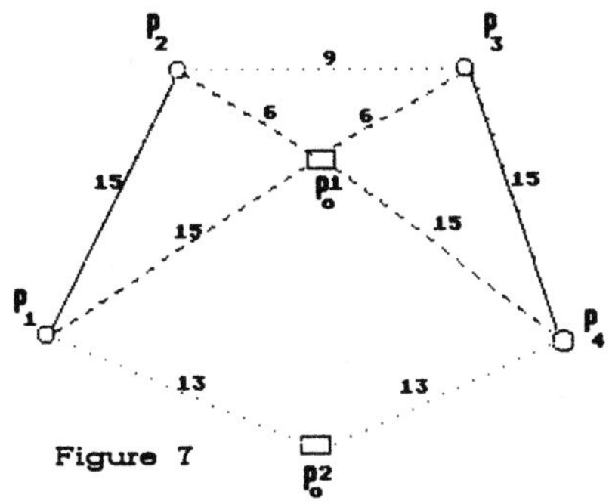

Figure 7

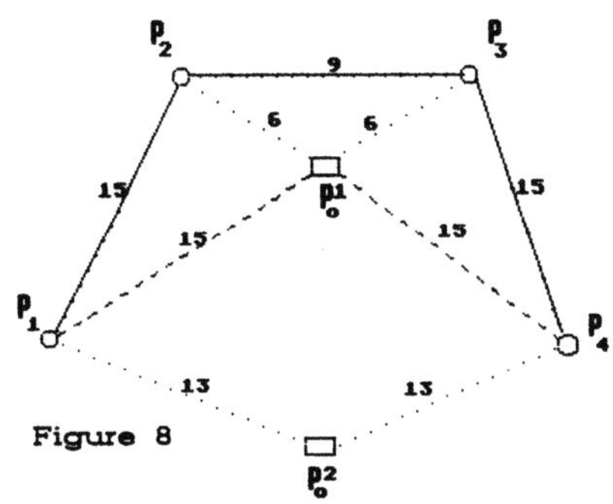

Figure 8

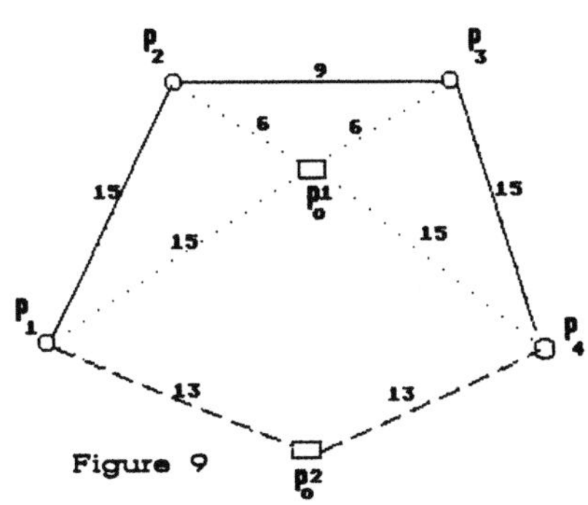

Figure 9

Bibliography

Bodin, L., B. Golden, A. Assad and M. Ball: Routing and Scheduling of Vehicles - the State of the Art, Comput. Oper. Res. 10 (1983) 2, p. 63 - 211

Clarke, G. and J. Wright: Scheduling of Vehicles from a Central Point to a Number of Delivery Points, Op. Res. 12 (1964) 4, p. 568 - 581

Fiala, F.: Vehicle Routing Problems, GMD-Mitteilungen No. 46 (1978)

Hausmann, D. and B. Korte: Lower Bounds on the Worst Case Complexity of some Oracle Algorithms, Discr. Math. 24 (1978), p. 261 - 276

Hausmann, D., B. Korte and T.A. Jenkins: Worst Case Analysis of Greedy Type Algorithms for Indepedent systems, Math. Prog. Study 12 (1980), p. 120 - 131

Terno, J.: Numerische Verfahren der diskreten Optimierung (Numerical Methods of Diskrete Optimization), Teubner Texte zur Mathematik No. 36 (1981)

2

Dynamic Systems

Enclosure Methods and Their Applications in Control Theory

Günter Ludyk

Abstract

Enclosure methods are described for the solution of linear and nonlinear systems of equations. These methods produce high precision solutions with the help of interval mathematics and the implemented optimal scalar product, e.g. in PASCAL-SC. Some problems of control theory are reformulated, so that they can be solved with the help of the enclosure methods. Especially, the problem of pole assignment is discussed in detail.

1. Introduction

In practical mathematics the new tool *interval mathematics* [1,2] has been developed, with which many engineering problems can be solved elegantly. In order to make full use of the advantages of interval mathematics with the help of computers, it is indispensible to have an optimal arithmetic. Two things are especially to emphasize: the possibility of rounding direction assignment and the high precision scalar product [3,4,5,15]. In the computer languages PASCAL-SC (PASCAL for Scientific Calculation) and FORTRAN-SC it is possible to specify rounding directions for the four elementary arithmetic operations $+, -, \cdot$ and $/$, namely, to round the result to the next larger or to the next smaller machine number. However, if in a computation process the differences of large numbers are to calculate, it is possible that the errors at the individual steps sum up to very large ones at the end. This can occur, e.g. in the calculation of scalar product which is an elementary operation for vector and matrix computation. Therefore, in the above languages a so-called *optimal scalar product* is introduced. For the realization of it, a long software accumulator is implemented, all intermediate results are stored in full length, and the rounding operation takes place only once at the end of the necessary calculations.

The in the above languages implemented arithmetic makes it possible to include the solutions of practically all basic problems of numerical mathematics in specified bounds with maximal precision. Furthermore, the solutions also can be verified with the arithmetic, i.e. it can be proven through an algorithm that there really exist solutions in the specified bounds and that they are unique.

The in section 2 described *enclosure methods* for the solution of linear and nonlinear systems of equations are suitable ones for these purposes [6,7,8]. All these procedures are iterative in nature, with which high precision solutions can be calculated in the form of *inclusion intervals* inspite of the finite precision of machine arithmetic. These inclusion *intervals* are generally so small, that their upper and lower bounds are different only at the last digit of the mantissa. Nevertheless, it is still always *guaranteed* that the exact solutions lie in the calculated intervals.

In section 3 and 4 some problems of control theory are reformulated, so that they can be solved with the help of the enclosure methods. Especially, the problem of pole assignment is discussed in detail [10].

2. Enclosure Methods for the Solution of Linear and Nonlinear Systems of Equations

a) The Iteration Method of NEWTON

The starting point of the enclosure methods is the iteration method of NEWTON for the calculation of zeros of differentiable functions $f(x)$. More precisely, this is a method for the determination of real numbers x for which there is $f(x) = 0$. This method comes of the well-known mean value theorem of differential calculus. The theorem states that for every differentiable function $f(x)$ there exists in any given interval $[a, b]$ at least one point x_0, at which the slope of the tangent line of $f(x)$ equals the slope of the line connecting the two end points $(a, f(a))$ and $(b, f(b))$. In other words, there is

$$f'(x_0) = \frac{f(b) - f(a)}{b - a}. \tag{1}$$

The equation of the tangent line is

$$y(x) = f(x_0) + f'(x_0)(x - x_0). \tag{2}$$

If the tangent line intersects the x-axis at a point x_1, there will be $y(x_1) = 0$ and the first approximation value of the zero of $f(x)$ is herewith obtained

$$x_1 = x_0 - [f'(x_0)]^{-1} f(x_0). \tag{3}$$

Now look at the tangent line of $f(x)$ at the point x_1. The slope of this line is $f'(x_1)$. Suppose it intersects the x-axis at $x = x_2$

$$x_2 = x_1 - [f'(x_1)]^{-1} f(x1). \tag{4}$$

This value can be seen as the second approximation of the zero of $f(x)$. The iteration method of NEWTON is obtained by continuing this process

$$x_{k+1} = x_k - [f'(x_k)]^{-1} f(x_k). \tag{5}$$

For a system of nonlinear functions $f_1, f_2, \cdots, f_n$, each of which is a function of n variables $x_1, x_2, \cdots, x_n$, the corresponding zero point problem can be formulated as follows: find the vector $\boldsymbol{x}$ which satisfies the following vector equation

$$\boldsymbol{f}(\boldsymbol{x}) = \mathbf{o} \tag{6}$$

where

$$\boldsymbol{x} \stackrel{\text{def}}{=} \begin{pmatrix} x_1 \\ \vdots \\ x_n \end{pmatrix} \quad \text{and} \quad \boldsymbol{f}(\boldsymbol{x}) \stackrel{\text{def}}{=} \begin{pmatrix} f_1(\boldsymbol{x}) \\ \vdots \\ f_n(\boldsymbol{x}) \end{pmatrix}. \tag{7}$$

Define the JACOBIAN as below

$$\boldsymbol{J}(\boldsymbol{x}) \stackrel{\text{def}}{=} \begin{pmatrix} \frac{\partial f_1}{\partial x_1} & \cdots & \frac{\partial f_1}{\partial x_n} \\ \vdots & & \vdots \\ \frac{\partial f_n}{\partial x_1} & \cdots & \frac{\partial f_n}{\partial x_n} \end{pmatrix}. \tag{8}$$

The basic iterative equation for the generalized NEWTON method reads

$$\boldsymbol{x}_{k+1} = \boldsymbol{x}_k - \boldsymbol{J}^{-1}(\boldsymbol{x}_k)\boldsymbol{f}(\boldsymbol{x}_k). \tag{9}$$

If, instead of the exact inverse matrix $\boldsymbol{J}^{-1}(\boldsymbol{x}_k)$, only an *approximation* $\boldsymbol{R}$ of it is used, the desired result can also be achieved with the algorithm below

$$\boldsymbol{x}_{k+1} = \boldsymbol{x}_k - \boldsymbol{R}\boldsymbol{f}(\boldsymbol{x}_k). \tag{10}$$

However, it should be pointed out in this case that, in order to guarantee the convergence of the approximation sequence, the eigenvalues of the difference matrix $\boldsymbol{I} - \boldsymbol{R}\boldsymbol{J}(\boldsymbol{x}_k)$ must have their absolute values less than one. See e.g. [9].

b) Exact Solutions of Systems of Linear Equations

The basic problem of linear algebra, i.e. the determination of the solution vector $\boldsymbol{x}$ for the following system of equations

$$\boldsymbol{A}\boldsymbol{x} = \boldsymbol{b} \tag{11}$$

can be reformulated to a zero point problem as below: Find the zero point $\boldsymbol{x}$ of the system of linear equations

$$\boldsymbol{f}(\boldsymbol{x}) = \boldsymbol{A}\boldsymbol{x} - \boldsymbol{b} = \mathbf{o}. \tag{12}$$

The JACOBIAN for this special case can be calculated easily through the partial differentiation of (11)

$$\boldsymbol{J} = \frac{\partial \boldsymbol{f}}{\partial \boldsymbol{x}} = \frac{\partial}{\partial \boldsymbol{x}}(\boldsymbol{A}\boldsymbol{x} - \boldsymbol{b}) = \boldsymbol{A}. \tag{13}$$

Substituting it in the equation (8), one obtains the following formula of iteration

$$\boldsymbol{x}_{k+1} = \boldsymbol{x}_k - \boldsymbol{A}^{-1}\boldsymbol{f}(\boldsymbol{x}_k). \tag{14}$$

If the inverse matrix $\boldsymbol{A}^{-1}$ is known exactly, the NEWTON method would furnish the exact solution after only one step:

$$\boldsymbol{x}_1 = \boldsymbol{x}_0 - \boldsymbol{A}^{-1}(\boldsymbol{A}\boldsymbol{x}_0 - \boldsymbol{b}) = \boldsymbol{A}^{-1}\boldsymbol{b}. \tag{15}$$

However, to obtain the exact inverse of a matrix is in general difficult. Accordingly, if an approximation $\boldsymbol{R}$ of the inverse matrix $\boldsymbol{A}^{-1}$ is used, one gets the following *simplified* NEWTON *method*

$$\boldsymbol{x}_{k+1} = \boldsymbol{x}_k - \boldsymbol{R}(\boldsymbol{A}\boldsymbol{x}_k - \boldsymbol{b}) \tag{16}$$

or

$$\boldsymbol{x}_{k+1} = \boldsymbol{R}\boldsymbol{b} + (\boldsymbol{I} - \boldsymbol{R}\boldsymbol{A})\boldsymbol{x}_k. \tag{17}$$

Define the error vector as

$$\tilde{\boldsymbol{x}} \stackrel{\text{def}}{=} \boldsymbol{A}\boldsymbol{x} - \boldsymbol{b}. \tag{18}$$

One gets the following relationships

$$\begin{aligned}\tilde{\boldsymbol{x}}_{k+1} &= \boldsymbol{A}\boldsymbol{x}_{k+1} - \boldsymbol{b} \\ &= \boldsymbol{A}(\boldsymbol{I} - \boldsymbol{R}\boldsymbol{A})\boldsymbol{x}_k + \boldsymbol{A}\boldsymbol{R}\boldsymbol{b} - \boldsymbol{b} \\ &= (\boldsymbol{I} - \boldsymbol{A}\boldsymbol{R})\boldsymbol{A}\boldsymbol{x}_k - (\boldsymbol{I} - \boldsymbol{A}\boldsymbol{R})\boldsymbol{b} \\ &= (\boldsymbol{I} - \boldsymbol{A}\boldsymbol{R})\tilde{\boldsymbol{x}}_k.\end{aligned}$$

Suppose that the starting error is $\tilde{\boldsymbol{x}}_0$. It is obvious from the last equation that after k steps the error becomes

$$\tilde{\boldsymbol{x}}_k = (\boldsymbol{I} - \boldsymbol{A}\boldsymbol{R})^k \tilde{\boldsymbol{x}}_0. \tag{19}$$

Provided that the eigenvalues of the difference matrix $\boldsymbol{I} - \boldsymbol{A}\boldsymbol{R}$ all have their absolute values less than one, the above error will go to zero as $k \to \infty$. Because $\boldsymbol{R} \approx \boldsymbol{A}^{-1}$, so that $\boldsymbol{I} - \boldsymbol{A}\boldsymbol{R} \approx \boldsymbol{O}$, which means that all the eigenvalues are close to zero and fast convergence can be expected.

Define an *interval vector* $[\boldsymbol{x}]_k$ to be such a vector, whose components are all intervals. If the vector $\boldsymbol{x}_k$ in (16) is replaced by the interval vector $[\boldsymbol{x}]_k$, the following interval vector equation is obtained

$$[\boldsymbol{x}]_{k+1} = \boldsymbol{R}\boldsymbol{b} + (\boldsymbol{I} - \boldsymbol{R}\boldsymbol{A})[\boldsymbol{x}]_k. \tag{20}$$

Now define

$$\boldsymbol{g}([\boldsymbol{x}]_k) \stackrel{\text{def}}{=} \boldsymbol{R}\boldsymbol{b} + (\boldsymbol{I} - \boldsymbol{R}\boldsymbol{A})[\boldsymbol{x}]_k. \tag{21}$$

From the SCHAUDER fixed point theorem it follows that $\boldsymbol{g}(\boldsymbol{x}) = \boldsymbol{x}$ at least has one solution $\boldsymbol{x}$ in $[\boldsymbol{x}]$ under the condition that

$$\boldsymbol{g}([\boldsymbol{x}]) \subseteq [\boldsymbol{x}]. \tag{22}$$

If such an interval vector $[\boldsymbol{x}]$ is computationally found so that the condition (22) is fulfilled, the *existence* of a solution $\boldsymbol{x}^* \in [\boldsymbol{x}]$ is proven herewith automatically. Furthermore, it is pointed out in [6] that such a solution $\boldsymbol{x}^*$ is even *unique*, if the following stricter condition is fulfilled

$$\boldsymbol{g}([\boldsymbol{x}]) \subseteq [\mathring{\boldsymbol{x}}] \tag{23}$$

where $[\mathring{\boldsymbol{x}}]$ denotes the *interior* of the interval vector $[\boldsymbol{x}]$.

c) Inclusion of the Defect of a System of Linear Equations

The solution intervals of a system of equations can be made further smaller, if the inclusion intervals are calculated for the *defect* $\boldsymbol{d}$ of the equation system rather than directly for the solution $\boldsymbol{x}$ of the system. The defect of an equation system is defined as

$$\boldsymbol{d} \stackrel{\text{def}}{=} \boldsymbol{A}\hat{\boldsymbol{x}} - \boldsymbol{b}, \tag{24}$$

where $\hat{\boldsymbol{x}}$ is an approximate solution. Denote the exact solution by $\boldsymbol{x}^*$, the following vector will be called the *solution error vector*

$$\bar{\boldsymbol{x}} \stackrel{\text{def}}{=} \boldsymbol{x}^* - \hat{\boldsymbol{x}} \tag{25}$$

Left multiplication of (25) by $\boldsymbol{A}$ and with $\boldsymbol{A}\boldsymbol{x}^* = \boldsymbol{b}$, one gets

$$\boldsymbol{A}\bar{\boldsymbol{x}} = \boldsymbol{b} - \boldsymbol{A}\hat{\boldsymbol{x}} = -\boldsymbol{d}. \tag{26}$$

Using (26) the following identity

$$\bar{\boldsymbol{x}} = -\boldsymbol{R}\boldsymbol{d} + \bar{\boldsymbol{x}} + \boldsymbol{R}\boldsymbol{d} \tag{27}$$

can be rewritten as

$$\bar{\boldsymbol{x}} = -\boldsymbol{R}\boldsymbol{d} + (\boldsymbol{I} - \boldsymbol{R}\boldsymbol{A})\bar{\boldsymbol{x}}. \tag{28}$$

If, analogous to (16), the following iterative algorithm is assumed

$$\bar{\boldsymbol{x}}_{k+1} := -\boldsymbol{R}\boldsymbol{d} + (\boldsymbol{I} - \boldsymbol{R}\boldsymbol{A})\bar{\boldsymbol{x}}_k \tag{29}$$

then, for the *defect error vector* $\boldsymbol{e}_k$,

$$\boldsymbol{e}_k \stackrel{\text{def}}{=} \boldsymbol{A}\bar{\boldsymbol{x}}_k - \boldsymbol{d} \tag{30}$$

the following *defect error equation* is obtained

$$\boldsymbol{e}_{k+1} = (\boldsymbol{I} - \boldsymbol{A}\boldsymbol{R})\boldsymbol{e}_k. \tag{31}$$

For an initial defect error vector $\boldsymbol{e}_0$ the iteration furnishes after k steps the following result

$$\boldsymbol{e}_k = (\boldsymbol{I} - \boldsymbol{A}\boldsymbol{R})^k \boldsymbol{e}_0, \tag{32}$$

Provided that the eigenvalues of the matrix $(\boldsymbol{I} - \boldsymbol{A}\boldsymbol{R})$ all have their absolute values less than one, the defect error vector will go to zero as $k \longrightarrow \infty$. If this is the case, the vector $\bar{\boldsymbol{x}}_k$ in (29) will approach the exact solution of $\boldsymbol{A}\bar{\boldsymbol{x}} = \boldsymbol{d}$, so that theoretically the exact solution $\boldsymbol{x}^*$ in (25) can be calculated with $\boldsymbol{x}^* = \hat{\boldsymbol{x}} + \bar{\boldsymbol{x}}$, where $\hat{\boldsymbol{x}}$ is an approximate solution obtained before. Practically this is impossible because of the rounding errors. Nevertheless, with the help of the following interval iteration formula

$$[\bar{\boldsymbol{x}}]_{k+1} := -\boldsymbol{R}\boldsymbol{d} + (\boldsymbol{I} - \boldsymbol{R}\boldsymbol{A})[\bar{\boldsymbol{x}}]_k \tag{33}$$

a high precision inclusion interval for $\bar{\boldsymbol{x}}$ can be calculated.

This leads to

Algorithm 1: {High Precision Computation of an Inclusion Interval $[\bar{\boldsymbol{x}}]$ for the Defect. The exact solution $\boldsymbol{x}^*$ is given by $\boldsymbol{x}^* \in \hat{\boldsymbol{x}} + [\bar{\boldsymbol{x}}]$}

1. Calculation of an approximate inverse $\boldsymbol{R}$ of $\boldsymbol{A}$, e.g. with the help of QR decomposition of $\boldsymbol{A}$;

2. Calculation of the following vectors with optimal scalar product:
   $\hat{\boldsymbol{x}} := \boldsymbol{R} * \boldsymbol{b}$; $\boldsymbol{d} := \boldsymbol{A} * \hat{\boldsymbol{x}} - \boldsymbol{b}$; $\boldsymbol{z} := -\boldsymbol{R} * \boldsymbol{d}$;

3. Calculation of $\boldsymbol{C} := \boldsymbol{I} - \boldsymbol{R} * \boldsymbol{A}$ with optimal scalar product;

4. Set $[\bar{\boldsymbol{x}}] := \boldsymbol{z}$ and $k := 0$;

5. **repeat**
   begin
   $[\boldsymbol{y}] := [\bar{\boldsymbol{x}}] \circ \epsilon$;
   $k := k + 1$;
   $[\bar{\boldsymbol{x}}] := (\boldsymbol{z} + \boldsymbol{C} * [\boldsymbol{y}]) \cap [\boldsymbol{y}]$;
   end;
   until $[\bar{\boldsymbol{x}}] = [\boldsymbol{y}]$ **or** $k = 10$;

6. **if** $[\bar{\boldsymbol{x}}] = [\boldsymbol{y}]$ **then**{there exists a unique solution $\boldsymbol{x}^*$ of $\boldsymbol{A}\boldsymbol{x} = \boldsymbol{b}$ and it is $\boldsymbol{x}^* \in \hat{\boldsymbol{x}} + [\bar{\boldsymbol{x}}]$}
else{it can not be determined, whether $\boldsymbol{A}$ is singular or not.}.

The operation $[\bar{\boldsymbol{x}}] \circ \epsilon$ at the 5th step of the algorithm denotes an enlargement of the interval vector $[\bar{\boldsymbol{x}}]$. This enlargement sometimes is necessary, in order to "capture" the real solution [8].

Implemented e.g. in PASCAL SC on a personal computer, the above algorithm will produce interval results whose upper and lower bounds are different only at the last digit of a mantissa of 12 digits, e.g.

$$x_i^* \in [2.124\,589\,763\,87E01, 2.124\,589\,763\,88E01]. \tag{34}$$

The algorithm 1 can also be applied in a modified form to such linear equation systems, whose matrix $\boldsymbol{A}$ and vector $\boldsymbol{b}$ have interval elements. The solution is also an interval vector. But in this case the lengths of the solution intervals depend also on the lengths of the intervals in the input data $\boldsymbol{A}$ and $\boldsymbol{b}$.

d) Inclusion of Solutions of Nonlinear Equation Systems

More presice results can be obtained also for *nonlinear* systems of equations $\boldsymbol{f}(\boldsymbol{x}) = \boldsymbol{o}$, if the interval iteration algorithm is carried out for the *defect* $\boldsymbol{d}$ of the nonlinear system of equations rather than directly for the solution $\boldsymbol{x}^*$. Suppose an approximate solution $\hat{\boldsymbol{x}}$ is already obtained. Similar to the case of linear equation systems, the defect of an nonlinear equation system is defined as

$$\boldsymbol{d} \stackrel{\text{def}}{=} \boldsymbol{f}(\hat{\boldsymbol{x}}). \tag{35}$$

Now if for the *solution error vector* $\bar{\boldsymbol{x}} = \boldsymbol{x}^* - \hat{\boldsymbol{x}}$ an inclusion interval vector $[\bar{\boldsymbol{x}}]$ is calculated, for the exact solution $\boldsymbol{x}^*$ there will hold:

$$\boldsymbol{x}^* \in \hat{\boldsymbol{x}} + [\bar{\boldsymbol{x}}] \tag{36}$$

Through iteration the largest interval length

$$\max_i w_i = \max_i (\bar{x}_{i,sup} - \bar{x}_{i,inf}) \tag{37}$$

can be made as small as possible. The terms $\bar{x}_{i,inf}$ and $\bar{x}_{i,sup}$ in (37) are the lower and upper bounds of the interval $[\bar{x}_i]$ respectively, i.e. $[\bar{x}_i] = [\bar{x}_{i,inf}, \bar{x}_{i,sup}]$.

The interval iteration formula for the calculation of a high precision inclusion for the solution error vector $\bar{\boldsymbol{x}}$ is similar to (33). The only difference is that now the JACOBI matrix $\boldsymbol{J}$ is used in place of $\boldsymbol{A}$

$$[\bar{\boldsymbol{x}}]_{k+1} := -\boldsymbol{R}\boldsymbol{d} + (\boldsymbol{I} - \boldsymbol{R}\boldsymbol{J})[\bar{\boldsymbol{x}}]_k, \tag{38}$$

The algorithm 1 is to change only slightly in this sense.

3. Pole Shifting

a) State Variable Feedback

With the help of state variable feedback, the dynamical performance of a control system can be changed arbitrarily. This is possible because the dynamical performance of a system is determined by its eigenvalues. The above statement is true for linear *continuous-time* systems with the following state variable equation

$$\dot{\boldsymbol{x}}(t) = \boldsymbol{A}\boldsymbol{x}(t) + \boldsymbol{b}u(t), \tag{39}$$

as well as for linear *discrete-time* systems described below:

$$\boldsymbol{x}_{k+1} = \boldsymbol{A}\boldsymbol{x}_k + \boldsymbol{b}u_k. \tag{40}$$

If the state variable vector $\boldsymbol{x}$ is fed back as follows

$$u = -\boldsymbol{r}^T\boldsymbol{x} + w, \tag{41}$$

a new system matrix $\boldsymbol{F}$ is obtained for both types of systems:

$$\boldsymbol{F} \stackrel{\text{def}}{=} (\boldsymbol{A} - \boldsymbol{b}\boldsymbol{r}^T). \tag{42}$$

Provided that the system is controllable, then the eigenvalues of $\boldsymbol{F}$ can be assigned arbitrarily through the feedback vector $\boldsymbol{r}$. Suppose the feedback system should have the following new characteristic polynomial

$$P(\lambda) = f_0 + f_1\lambda + f_2\lambda^2 + \cdots + f_{n-1}\lambda^{n-1} + \lambda^n \tag{43}$$

the corresponding feedback vector $\boldsymbol{r}$ can be calculated, e.g. using the so-called formula of ACKERMANN [11]

$$\boldsymbol{r}^T = \boldsymbol{q}^T P(\boldsymbol{A}) \tag{44}$$

where the row vector $\boldsymbol{q}^T$ is equal to the last row of the inverse of the controllability matrix

$$\boldsymbol{S} = [\boldsymbol{b}, \boldsymbol{A}\boldsymbol{b}, \ldots, \boldsymbol{A}^{n-1}\boldsymbol{b}], \tag{45}$$

or in other words

$$\boldsymbol{q}^T = [0, \ldots, 0, 1] \cdot \boldsymbol{S}^{-1}, \tag{46}$$

and P($\boldsymbol{A}$) is such a matrix which results from substituting the matrix $\boldsymbol{A}$ for λ in the characteristic polynomial

$$P(\boldsymbol{A}) = f_0\boldsymbol{I} + f_1\boldsymbol{A} + f_2\boldsymbol{A}^2 + \cdots + f_{n-1}\boldsymbol{A}^{n-1} + \boldsymbol{A}^n. \tag{47}$$

b) High Precision Computation of the Controllability Matrix and the Vector $\boldsymbol{q}$

The controllability matrix $\boldsymbol{S}$ has the property, that with growing i its column vectors $\boldsymbol{A}^i\boldsymbol{b}$ show the tendancy to converge to the eigenvector belonging to the dominant eigenvalue. In other words, the column vectors of $\boldsymbol{S}$ become more and more linearly dependant on each other with growing i, see e.g. [9]. In this way, the controllability matrix often is badly conditioned. Consequently, the use of ACKERMANN's formula is warned up to now for calculation of the feedback vector $\boldsymbol{r}$, see [12]. However, these difficulties can be overcome with the help of inclusion procedures introduced in this paper. For this purpose, it is necessary to reformulate the task at first as several linear or nonlinear zero point problems.

The computation of the controllability matrix can be reduced to the problem of solving a system of linear equations as follows. Create a linear equation system as below:

$$\begin{pmatrix} \boldsymbol{I} & & & & \boldsymbol{O} \\ -\boldsymbol{A} & \boldsymbol{I} & & & \\ & -\boldsymbol{A} & \ddots & & \\ & & \ddots & \boldsymbol{I} & \\ \boldsymbol{O} & & & -\boldsymbol{A} & \boldsymbol{I} \end{pmatrix} \begin{pmatrix} \boldsymbol{s}_1 \\ \boldsymbol{s}_2 \\ \vdots \\ \boldsymbol{s}_n \end{pmatrix} = \begin{pmatrix} \boldsymbol{b} \\ \boldsymbol{o} \\ \vdots \\ \boldsymbol{o} \end{pmatrix}, \tag{48}$$

it can be seen from it that

$$\begin{aligned} \boldsymbol{s}_1 &= \boldsymbol{b}, \\ -\boldsymbol{A}\boldsymbol{s}_1 + \boldsymbol{s}_2 &= \boldsymbol{o}, \;\rightarrow\; \boldsymbol{s}_1 = \boldsymbol{A}\boldsymbol{b}, \\ -\boldsymbol{A}\boldsymbol{s}_2 + \boldsymbol{s}_3 &= \boldsymbol{o}, \;\rightarrow\; \boldsymbol{s}_3 = \boldsymbol{A}^2\boldsymbol{b} \end{aligned} \tag{49}$$

and so on up to

$$\boldsymbol{s}_n = \boldsymbol{A}^{n-1}\boldsymbol{b}.$$

The above relations mean that the controllability matrix $\boldsymbol{S}$ is composed of the subvectors $\boldsymbol{s}_i$ of the solution vector of (48), namely

$$\boldsymbol{S} = [\boldsymbol{s}_1, \ldots, \boldsymbol{s}_n], \tag{50}$$

so that it can be calculated with high precision using algorithm 1. Then the so obtained controllability matrix $\boldsymbol{S}$ is an interval matrix, whose each element is an interval of the form (34).

For the calculation of the vector $\boldsymbol{q}$ according to (45), the computation of the inverse of the controllability matrix $\boldsymbol{S}^{-1}$ can be avoided as follows: with

$$\boldsymbol{i}_n^T \stackrel{\text{def}}{=} [0, \ldots, 0, 1] \tag{51}$$

the formula

$$\boldsymbol{q}^T = \boldsymbol{i}_n^T \boldsymbol{S}^{-1} \tag{52}$$

can be transformed to a system of linear equations for $\boldsymbol{q}$ as below through right multiplication by $\boldsymbol{S}$ and transposition

$$\boldsymbol{S}^T \boldsymbol{q} = \boldsymbol{i}_n. \tag{53}$$

Note that the elements of the controllability matrix are now all intervals, so that the calculated solution vector $\boldsymbol{q}$ will have interval components, whose lengths can no more be made as small as possible. They are now dependant on the interval lengths in $\boldsymbol{S}$.

c) Calculation of the Feedback Vector $\boldsymbol{r}$

Combining the relations (47) and (44) and transposing the obtained equation, one gets the following formula for calculation of the feedback vector

$$\boldsymbol{r} = f_0\boldsymbol{q} + f_1\boldsymbol{A}^T\boldsymbol{q} + \cdots + f_{n-1}(\boldsymbol{A}^T)^{n-1}\boldsymbol{q} + (\boldsymbol{A}^T)^n\boldsymbol{q}. \tag{54}$$

Create a new linear equation system

$$\begin{pmatrix} \boldsymbol{I} & & & \boldsymbol{O} \\ -\boldsymbol{A}^T & \boldsymbol{I} & & \\ & \ddots & \ddots & \\ \boldsymbol{O} & & -\boldsymbol{A}^T & \boldsymbol{I} \end{pmatrix} \begin{pmatrix} \boldsymbol{x}_1 \\ \boldsymbol{x}_2 \\ \vdots \\ \boldsymbol{x}_{n+1} \end{pmatrix} = \begin{pmatrix} \boldsymbol{q} \\ f_{n-1}\boldsymbol{q} \\ \vdots \\ f_0\boldsymbol{q} \end{pmatrix}, \tag{55}$$

it can be seen that

$$\begin{array}{rcl} \boldsymbol{x}_1 & = & \boldsymbol{q}; \\ -\boldsymbol{A}^T\boldsymbol{x}_1 + \boldsymbol{x}_2 = f_{n-1}\boldsymbol{q}, & \rightarrow & \boldsymbol{x}_2 = f_{n-1}\boldsymbol{q} + \boldsymbol{A}^T\boldsymbol{q}; \\ -\boldsymbol{A}^T\boldsymbol{x}_2 + \boldsymbol{x}_3 = f_{n-2}\boldsymbol{q}, & \rightarrow & \boldsymbol{x}_3 = f_{n-2}\boldsymbol{q} + f_{n-1}\boldsymbol{A}^T\boldsymbol{q} + (\boldsymbol{A}^T)^2\boldsymbol{q}; \\ & \vdots & \end{array}$$

and finally

$$\boldsymbol{x}_{n+1} = f_0\boldsymbol{q} + f_1\boldsymbol{A}^T\boldsymbol{q} + \cdots + f_{n-1}(\boldsymbol{A}^T)^{n-1}\boldsymbol{q} + (\boldsymbol{A}^T)^n\boldsymbol{q}, \tag{56}$$

i.e. there is directly

$$\boldsymbol{r} = \boldsymbol{x}_{n+1}. \tag{57}$$

Since the vector $\boldsymbol{q}$ is the solution vector of the equation system (53), the right side of the equation system (55) is also an interval vector. Therefore, the solution vector of the equation system (55) and accordingly the feedback vector $\boldsymbol{r}$ will have greater interval lengths than the vector $\boldsymbol{q}$. In other words, it becomes less precise.

d) High Precision Calculation of the Feedback Vector $\boldsymbol{r}$

In the above, the computation is carried out in 3 steps. At the first step the controllability matrix $\boldsymbol{S}$ is calculated in form of an interval matrix $[\boldsymbol{S}]$, whose elements are precise until the last digit of the mantissa. At the second step, the interval vector $[\boldsymbol{q}]$ is computed using the transposed interval matrix $[\boldsymbol{S}]^T$. Here the computed intervals can have bigger lengths, if the controllability matrix $\boldsymbol{S}$ is badly conditioned. The feedback interval vector $[\boldsymbol{r}]$ is calculated at the third step using the computed interval vector $[\boldsymbol{q}]$. Now, it should be pointed out that although the calculated interval vector $[\boldsymbol{r}]$ includes in it the exact solution, but the interval lengths of its components can be so large that the result has no use any more.

In order to overcome this difficulty, the three equation systems (48), (53) and (55) can be written together as a new one:

$$\begin{pmatrix} I & & & & & & & & O \\ -A & I & & & & & & & \\ & \ddots & \ddots & & & & & & \\ & & -A & I & & & & & \\ & & & O & S^T & & & & \\ & & & & -A^T & I & & & \\ & & & & & -A^T & I & & \\ & & & & & & \ddots & \ddots & \\ O & & & & & & & -A^T & I \end{pmatrix} \begin{pmatrix} s_1 \\ s_2 \\ \vdots \\ s_n \\ q \\ x_2 \\ x_3 \\ \vdots \\ r \end{pmatrix} = \begin{pmatrix} b \\ o \\ \vdots \\ o \\ i_n \\ f_{n-1}q \\ f_{n-2}q \\ \vdots \\ f_0 q \end{pmatrix} \tag{58}$$

Note that this new equation system is *no more* a linear equation system, because there exist unknown variables not only at the left side of the equation in S^T, but also at the right side in q. Move the vector at the right side of (58) to the left side and develop the terms, one gets the following *nonlinear* equation system

$$\begin{pmatrix} s_1 - b \\ s_2 - As_1 \\ \vdots \\ s_n - As_{n-1} \\ s_1^T q \\ s_2^T q \\ \vdots \\ s_n^T q - 1 \\ x_2 - A^T q - f_{n-1} q \\ x_3 - A^T x_2 - f_{n-2} q \\ \vdots \\ r - A^T x_n - f_0 q \end{pmatrix} = o. \tag{59}$$

This system of nonlinear equations can now be solved with high precision with the help of the in section 2d) described enclosure methods. Here it should be pointed out that the feedback vector r can now be calculated also with high precision, i.e. as in (34), despite of the bad condition of the matrix A. The approximate solution $\hat{x}$ needed by the inclusion algorithm can be obtained by taking the average of the more or less precise solutions of the equation systems (48), (53) and (55).

The above described method is generalized to the pole shifting problem for multivariable systems in [9].

e) Example

Given a linear system of second order with the following system matrix and input vector

$$A = \begin{pmatrix} 1 & \epsilon \\ 0 & 1 \end{pmatrix} \quad \text{und} \quad b = \begin{pmatrix} 1 \\ \epsilon \end{pmatrix}.$$

For $\epsilon = 10^{-3}, 10^{-4}$ and 10^{-5} respectively, calculate a feedback vector r, so that the system matrix eigenvalues $\lambda_1 = \lambda_2 = 1$ be shifted to $\lambda_1 = -1$ and $\lambda_2 = -2$. In other words, calculate $\boldsymbol{r}$ such that the characteristic polynomial of the system becomes

$$(\lambda-\lambda_1)(\lambda-\lambda_2)=(\lambda+1)(\lambda+2)=2+3\lambda+\lambda^2=f_0+f_1\lambda+\lambda^2.$$

Applying e.g. the in MATLAB [12] implemented procedures "acker" and "place" to this problem, one gets for the case $\epsilon=10^{-3}$ the following two feedback vectors

$$\boldsymbol{r}_{acker}=\begin{pmatrix}6.000\,000\,000\,493\,793E+006\\-5.999\,995\,000\,493\,793E+009\end{pmatrix}$$

and

$$\boldsymbol{r}_{place}=\begin{pmatrix}5.999\,988\,505\,986\,243E+006\\-5.999\,983\,505\,990\,704E+009\end{pmatrix}.$$

Checking the achieved system matrix eigenvalues with these two feedback vectors, one gets using MATLAB

$$\begin{aligned}\lambda_{1,acker}&=-1.995\,689\,888\,484\,776,\\\lambda_{2,acker}&=-1.004\,310\,111\,514\,769\end{aligned}$$

and

$$\begin{aligned}\lambda_{1,place}&=-1.996\,383\,633\,464\,575,\\\lambda_{2,place}&=-1.003\,611\,905\,965\,954.\end{aligned}$$

In the case $\epsilon=10^{-4}$ one gets using MATLAB

$$\boldsymbol{r}_{acker}=\begin{pmatrix}6.000\,000\,036\,481\,788E+008\\-5.999\,999\,986\,481\,788E+012\end{pmatrix}$$

and

$$\boldsymbol{r}_{place}=\begin{pmatrix}5.996\,803\,784\,804\,430E+008\\-5.996\,803\,734\,823\,565E+012\end{pmatrix}.$$

Eigenvalue checking gives

$$\begin{aligned}\lambda_{1,acker}&=-1.500\,000\,000\,000\,000+3.414\,125\,070\,936\,916\,i\\\lambda_{2,acker}&=-1.500\,000\,000\,000\,000-3.414\,125\,070\,936\,916\,i\end{aligned}$$

and

$$\begin{aligned}\lambda_{1,place}&=-1.499\,043\,226\,242\,065+5.326\,818\,938\,165\,629\,i,\\\lambda_{2,place}&=-1.499\,043\,226\,242\,065-5.326\,818\,938\,165\,629\,i.\end{aligned}$$

Finally, in the case $\epsilon=10^{-5}$, the "acker" procedure in MATLAB supplies

$$\boldsymbol{r}_{acker}=\begin{pmatrix}5.999\,999\,501\,779\,047E+010\\-5.999\,999\,501\,279\,046E+015\end{pmatrix},$$

with the achieved eigenvalues

$$\lambda_{1,acker}=-1.500\,000\,000\,000\,000+6.704\,744\,588\,722\,228E+002\,i,$$

$$\lambda_{2,acker} \quad = \quad -1.500\,000\,000\,000\,000 - 6.704\,744\,588\,722\,228E + 002\,i,$$

Note that instead of the desired real eigenvalues −1 and −2 one gets now a pair of conjugated complex poles having an imaginary part of three integral digits ! The other procedure "place" is even aborted for this case with the error message: "can't place eigenvalues there" !

In contrast, table 1 in the appendix is calculated for the feedback vector using the inclusion procedures described above. The symbol $\boldsymbol{r}_{exact}$ in the table denotes the exact values which are easy to calculate for this example. It can be seen that the calculated inclusion intervals are precise until the last digit of the 12 digit mantissa.

A popular checking method for computational algorithms for calculation of feedback vectors is that the original eigenvalues of the system matrix $\boldsymbol{A}$ are given back as new desired eigenvalues to the controlled system [10]. The exact feedback vector $\boldsymbol{r}$ now should be naturally zero. In this example, the "new" eigenvalues $\lambda_1 = \lambda_2 = 1$ are given back, and the inclusion algorithm produced the following results

$$\boldsymbol{r} = \begin{pmatrix} [-1.0 \cdot 10^{-99}, +1.0 \cdot 10^{-99}] \\ [-1.0 \cdot 10^{-99}, +1.0 \cdot 10^{-99}] \end{pmatrix},$$

where $1.0 \cdot 10^{-99}$ is the smallest machine number different from zero in the used language PASCAL-SC. This means, the zero vector is obtained as precisely as possible in the inclusion vector!

4. Other Applications of Enclosure Methods

In the following, several other problems from system and control theory are discussed. These problems can all be reduced to the problem of solving nonlinear equation systems, so that the high precision inclusion algorithms can be applied.

a) Calculation of Eigenvalues and Eigenvectors

Between an eigenvalue λ_i and its belonging eigenvector $\boldsymbol{x}_i$ of a matrix $\boldsymbol{A}$ there holds:

$$\boldsymbol{A}\boldsymbol{x}_i = \lambda_i \boldsymbol{x}_i. \tag{60}$$

These are n equations for the eigenvector itself and the n components of it, namely, there are altogether $n+1$ unknown variables. Because the length of the eigenvector is not fixed, a scaling equation for the eigenvector can be added to them as the $(n+1)$-th equation. If it is determined through e.g. approximation methods that the j-th component $x_{i,j}$ of the eigenvector has the largest absolute value, then this value can be scaled as one. Denote by $\boldsymbol{i}_j$ the j-th unit vector

$$\boldsymbol{i}_j = [0, \ldots, 0, 1, 0, \ldots, 0], \tag{61}$$

where only the j-th component is one. The scaling equation can now be written as

$$\boldsymbol{x}_i^T \boldsymbol{i}_j = 1. \tag{62}$$

Combining (60) and (62), the following system of nonlinear equations is obtained:

$$f(\lambda_i, x_i) = \begin{pmatrix} (\lambda_i I - A)x_i \\ x_i^T i_j - 1 \end{pmatrix} = o. \tag{63}$$

The zero vector of the above equation system, which consists of λ_i and x_i, can be calculated with high precision using the inclusion algorithms. In [9] and [13] respectively is described how to tackle the case of complex eigenvalues and the case of multiple eigenvalues.

b) Singular Value Decomposition

For any given $m \times n$ matrix A, there always exist two orthogonal matrices U and V, such that

$$U^T A V = \begin{pmatrix} \Sigma \\ O \end{pmatrix}, \tag{64}$$

where Σ is a diagonal matrix

$$\Sigma = \begin{pmatrix} \sigma_1 & & & \\ & \sigma_2 & & \\ & & \ddots & \\ & & & \sigma_n \end{pmatrix} \tag{65}$$

and there is

$$\sigma_1 \geq \sigma_2 \geq \cdots \geq \sigma_n \geq 0. \tag{66}$$

The n numbers $\sigma_1, \sigma_2, \cdots, \sigma_n$ are called the *singular values* of the matrix A. The singular values of a matrix can be used easily to give information about the rank, the norm and the condition number of the matrix. And the orthogonal matrices U and V can help in obtaining basis vectors for certain subspaces which are spanned by the column vectors of the matrix A.

If, for example, the rank of a matrix is to be determined with the help of its singular values, it is then necessary to know how many singular values are non-zero. The usual algorithms for calculation of singular values will fail for this task, especially when the given matrix is badly conditioned. Here the smallest singular value is to be determined as precisely as possible. In this case the guaranteed interval inclusion is more desirable. Because if an inclusion interval does not contain zero, it is then guaranteed that the corresponding singular value is non-zero.

The high precision calculation of inclusion intervals for singular values and their belonging singular vectors u_i and v_i, which are the columns of the matrices U and V in (64), can be reduced to the calculation of inclusion intervals for eigenvalues and eigenvectors, i.e. once more to the problem of solving nonlinear equation systems.

A practical algorithm for the calculation can be created with the help of the following propositions, for whose proofs see e.g. [9]:

1. The non-negative eigenvalues λ_i of the block matrix

$$A^* \stackrel{\text{def}}{=} \begin{pmatrix} O & A^T \\ A & O \end{pmatrix} \tag{67}$$

are equal to the singular values of the matrix A.

2. The belonging eigenvector $\boldsymbol{x}_i$ of the matrix $\boldsymbol{A}^*$ consists of the singular vectors $\boldsymbol{u}_i$ and $\boldsymbol{v}_i$ in the form

$$\boldsymbol{x}_i = \begin{pmatrix} \boldsymbol{u}_i \\ \boldsymbol{v}_i \end{pmatrix}. \tag{68}$$

3. There holds furthermore

$$\boldsymbol{u}_i^T \boldsymbol{u}_i = \boldsymbol{v}_i^T \boldsymbol{v}_i. \tag{69}$$

The following nonlinear equation system

$$\boldsymbol{f}(\sigma_i, \boldsymbol{x}_i) = \begin{pmatrix} (\sigma_i \boldsymbol{I} - \boldsymbol{A}^*)\boldsymbol{x}_i \\ \boldsymbol{x}_i^T \boldsymbol{x}_i - 2 \end{pmatrix} = \mathbf{o} \tag{70}$$

has a zero point, which consists of σ_i and $\boldsymbol{x}_i$. This zero point satisfies naturally the last row in the equation system, which is

$$\boldsymbol{x}_i^T \boldsymbol{x}_i = 2 = \boldsymbol{u}_i^T \boldsymbol{u}_i + \boldsymbol{v}_i^T \boldsymbol{v}_i. \tag{71}$$

It follows from (69) that

$$\boldsymbol{u}_i^T \boldsymbol{u}_i = \boldsymbol{v}_i^T \boldsymbol{v}_i = 1. \tag{72}$$

This is a necessary condition for the orthonormal singular vectors. Substitute $\boldsymbol{A}^*$ of (67) and $\boldsymbol{x}_i$ of (68) into (70), a system of nonlinear equations is obtained as below which can be solved now directly

$$\boldsymbol{f}(\sigma_i, \boldsymbol{u}_i, \boldsymbol{v}_i) = \begin{pmatrix} \sigma_i \boldsymbol{u}_i - \boldsymbol{A}^T \boldsymbol{v}_i \\ \sigma_i \boldsymbol{v}_i - \boldsymbol{A}^T \boldsymbol{u}_i \\ \boldsymbol{u}_i^T \boldsymbol{u}_i + \boldsymbol{v}_i^T \boldsymbol{v}_i - 2 \end{pmatrix} = \mathbf{o}. \tag{73}$$

In [9,16] are described many other applications of the technique of singular value decomposition, such as

- transformation of a mathematical description of a multivariable system to block-HESSENBERG-form,
- canonical system decomposition of KALMAN,
- geometrical theory of disturbance decoupling,
- balanced realisation and model reduction,
- least square methods and pseudoinverse methods.

c) Simulation of Nonlinear Systems

At the first look, there seems to be no difficulties in the simulation of nonlinear discrete systems

$$\boldsymbol{x}_{k+1} = \boldsymbol{f}(\boldsymbol{x}_k), \tag{74}$$

since it seems that the calculation can be carried out from an initial state $\boldsymbol{x}_0$ smoothly in the sequence $\boldsymbol{x}_1 = \boldsymbol{f}(\boldsymbol{x}_0)$, $\boldsymbol{x}_2 = \boldsymbol{f}(\boldsymbol{x}_1)$ and so on. However, if one takes the following example

$$x_{k+1} = x_k^2 - 1.7 \tag{75}$$

starting from e.g. the initial state $x_0 = 1$, one will get after roughly $k = 70$ steps only pure stochastic numbers, which results from the accumulating rounding errors and which has to do nothing more with the *chaotic* but real system dynamics [14].

The best numerical method, with which the states x_k for large k values can be calculated with maximal precision, is the enclosure method. For example, if starting from an initial state x_0 a hundred states x_1 to x_{100} are to be calculated for the nonlinear system (75), one can create a 100 dimensional vector $\boldsymbol{x}$ with the hundred states as its components, and gets the following nonlinear equation system

$$\boldsymbol{x} = \boldsymbol{g}(\boldsymbol{x}) \tag{76}$$

with

$$\boldsymbol{x} \stackrel{\text{def}}{=} \begin{pmatrix} x_1 \\ x_2 \\ \vdots \\ x_{100} \end{pmatrix} \quad \text{and} \quad \boldsymbol{g}(\boldsymbol{x}) = \begin{pmatrix} x_0^2 - 1.7 \\ x_1^2 - 1.7 \\ \vdots \\ x_{100}^2 - 1.7 \end{pmatrix}.$$

Reformulate equation (76) as

$$\boldsymbol{f}(\boldsymbol{x}) \stackrel{\text{def}}{=} \boldsymbol{x} - \boldsymbol{g}(\boldsymbol{x}), \tag{77}$$

one gets once more the "zero point problem", which can be solved with high precision using the inclusion procedures described in this paper.

References

[1] Alefeld,G., Herzberger, J.: Introduction to interval computations. Academic Press, New York 1983.

[2] Bauch,H., Jahn,K.-U., Oelschlägel,D., Süsse,H., Wiebig,V.: Intervallmathematik. Teubner, Leipzig 1987.

[3] Kulisch,U.: Eine neue Arithmetik für wissenschaftliches Rechnen. In: Kulisch,U., Ullrich,Ch.(editor): Wissenschaftliches Rechnen und Programmiersprachen. Teubner, Stuttgart 1982.

[4] Bohlender,G., et al.: PASCAL-SC. Bibliographisches Institut, Mannheim 1986.

[5] Kießling,I., Lowes,M., Paulik,A.: Genaue Rechnerarithmetik. Intervallrechnung und Programmieren mit PASCAL-SC. Teubner, Stuttgart 1988.

[6] Rump,S.M.: Solving Algebraic Problems with High Accuracy. In: Kulisch,U., Miranker,W.L.(editor): A New Approach to Scientific Computation. Academic Press, New York 1981.

[7] Rump,S.M.: New Results on Verified Inclusions. In: Miranker,L.,Toupin,R., (editor): Accurate Scientific Computation. Springer, Berlin 1986.

[8] Böhm,H.,Rump,S.M.,Schumacher,G.: E-Methods for Nonlinear Problems. In: Kaucher,E.,Kulisch,U.,Ullrich,Ch.(editor): Computerarithmetic. Teubner, Stuttgart 1987.

[9] Ludyk,G.:CAE von Dynamischen Systemen: Analyse, Simulation, Entwurf von Regelungssystemen. Springer, Berlin 1990.

[10] Miminis,G.S.,Paige,C.C.: A Direct Algorithm for Pole Assignment of Time-invariant Multi-input Linear Systems using State Feedback. Automatica 24(1988)343-356.

[11] Ackermann,J.:Der Entwurf linearer Regelungssysteme im Zustandsraum. Regelungstechnik 7(1972)297-300.

[12] Laub,A.J.,Little,J.N.: Control System Toolbox for use with MATLAB. The Math Works Inc., Sherborn MA 1986.

[13] Alefeld,G.,Spreuer,H.: Iterative Improvement of Componentwise Errorbounds for Invariant Subspaces Belonging to a Double or Nearly Double Eigenvalue. Computing 36(1986)321-334.

[14] Ludyk,G.: Nichtlineare zeitdiskrete Systeme. Automatisierungstechnik 36(1988)321-330.

[15] Kulisch,U.(editor): PASCAL-SC, a PASCAL Extension for Scientific Computation. Teubner, Stuttgart, und John Wiley, Chichester 1987.

[16] Klema,V.C., Laub,A.J.: The singular value decomposition: Its computation and some applications. IEEE Trans. Autom. Control, 25(1980)164-176.

Table 1

ϵ	r_{exact}	$[r]_{calculated}$
10^{-3}	$r_1 = 6 \cdot 10^6$ $r_2 = -5.999\,995\,000\,00 \cdot 10^9$	$[r_1] = [5.999\,999\,999\,99 \cdot 10^6, 6.000\,000\,000\,01 \cdot 10^6]$ $[r_2] = [-5.999\,995\,000\,01 \cdot 10^9, -5.999\,994\,999\,99 \cdot 10^9]$
10^{-4}	$r_1 = 6 \cdot 10^8$ $r_2 = -5.999\,999\,950\,00 \cdot 10^{12}$	$[r_1] = [5.999\,999\,999\,99 \cdot 10^8, 6.000\,000\,000\,01 \cdot 10^8]$ $[r_2] = [-5.999\,999\,950\,01 \cdot 10^{12}, -5.999\,999\,949\,99 \cdot 10^{12}]$
10^{-5}	$r_1 = 6 \cdot 10^{10}$ $r_2 = -5.999\,999\,999\,50 \cdot 10^{15}$	$[r_1] = [5.999\,999\,999\,99 \cdot 10^{10}, 6.000\,000\,000\,01 \cdot 10^{10}]$ $[r_2] = [-5.999\,999\,999\,51 \cdot 10^{15}, -5.999\,999\,999\,49 \cdot 10^{15}]$

Improved Computation of Balancing Transformations for Model Reduction of Minimal Systems

H.- W. Philippsen

Abstract

The computation of balancing transformations by the aid of the *classical* algorithm of Laub [6] is feasible for very good conditioned problems only. Numerical problems arise in several matrix products, the computation of an orthogonal transformation and in solving Lyapunov equations. Laub et al [7] made investigations to overcome these problems by means of a short and elegant algorithm. Backbone of their method is an algorithm proposed by Hammarling [3] calculating required Cholesky factors directly out of the Lyapunov equations. This paper will present a new improved method computing the Cholesky factors of the Lyapunov equation via nonlinear equations. The solution is highly accurate and can be verified by interval arithmetic. Additionally the optimal scalar product [5] will be proposed for the high accurate computation of matrix products. The presented methods will improve the computation of balancing transformation of minimal systems. They are a first step towards results with guaranteed precision in the computation of control systems.

1 Introduction

A balancing transformation is a similarity transformation of a linear time invariant not necessarily continuous or minimal system:

$$\begin{aligned} \dot{\mathbf{x}}(t) &= \mathbf{A}\mathbf{x}(t) + \mathbf{B}\mathbf{u}(t) \\ \mathbf{y}(t) &= \mathbf{C}\mathbf{x}(t) \end{aligned} \tag{1}$$

where $\mathbf{A} \in \mathbf{R}^{n \times n}$, $\mathbf{B} \in \mathbf{R}^{n \times m}$, $\mathbf{C} \in \mathbf{R}^{p \times n}$ and

$$\mathbf{A}_{Bal} = \mathbf{T}_{Bal}\mathbf{A}\mathbf{T}_{Bal}^{-1} \; , \; \mathbf{B}_{Bal} = \mathbf{T}_{Bal}\mathbf{B} \; , \; \mathbf{C}_{Bal} = \mathbf{C}\mathbf{T}_{Bal}^{-1} \tag{2}$$

where $\mathbf{T_{Bal}} \in \mathbf{R}^{n \times n}$. The corresponding reachebility Gramian of the system (1)

$$\mathbf{P} = \int_0^\infty e^{\mathbf{A}t}\mathbf{B}\mathbf{B}^\top e^{\mathbf{A}^\top t} dt \tag{3}$$

and the observability Gramian

$$\mathbf{Q} = \int_0^\infty e^{\mathbf{A}^\top t}\mathbf{C}^\top \mathbf{C} e^{\mathbf{A}t} dt \tag{4}$$

are transformed by a similarity transformation (2):

$$\mathbf{P}_{Bal} = \mathbf{T}_{Bal}\mathbf{P}\mathbf{T}_{Bal}^{\mathsf{T}} \tag{5}$$

$$\mathbf{Q}_{Bal} = \mathbf{T}_{Bal}^{-1^{\mathsf{T}}}\mathbf{Q}\mathbf{T}_{Bal}^{-1} \tag{6}$$

It can be easily checked that a similarity transformation of the system is a similarity transformation of the matrix product **PQ**. So the eigenvalues of **PQ** are invariant under similarity transformation. It can be shown [9], [2], that there exists a transformation (2) such that the Gram's have the form

$$\mathbf{P}_{Bal} = \mathbf{Q}_{Bal} = \mathbf{\Sigma} = diag(\sigma_1, \ldots, \sigma_n) \tag{7}$$

$$\sigma_1 \geq \sigma_2 \ldots \geq \sigma_n > 0$$

where $\sigma_i \in \mathbf{R}$ are the so called singular values of the system. The σ_i are defined by

$$\sigma_i = \sqrt{\lambda_i(\mathbf{PQ})} \tag{8}$$

where λ_i denotes the i th eigenvalue of **PQ**. The singular values play an important role for model reduction [2] not only for balanced systems [10]. The σ_i give error bounds in the sense of the L^∞– Norm of the difference frequency response of original and reduced model [2].

P and **Q** are calculated via the Lyapunov equations:

$$\mathbf{AP} + \mathbf{PA}^{\mathsf{T}} = -\mathbf{BB}^{\mathsf{T}} \tag{9}$$

$$\mathbf{A}^{\mathsf{T}}\mathbf{Q} + \mathbf{QA} = -\mathbf{C}^{\mathsf{T}}\mathbf{C} \tag{10}$$

In the case of minimal system $\mathbf{P}, \mathbf{Q}$ are positive definite, so real Cholesky factors can be used computing right and left eigenspaces of **PQ**. The improved method of Laub et al [7] is executed in four main steps:

B 1 : Calculate Cholesky factors of the Gramians directly out of the Lyapunov equations (9) and (10)

$$\mathbf{P} = \mathbf{R}_P^{\mathsf{T}}\mathbf{R}_P \ , \ \mathbf{Q} = \mathbf{R}_Q^{\mathsf{T}}\mathbf{R}_Q$$

B 2 : Compute the singular value decomposition of

$$\mathbf{R}_Q\mathbf{R}_P^{\mathsf{T}} = \mathbf{U}\mathbf{\Sigma}\mathbf{V}^{\mathsf{T}}$$

B 3 : Form the transformation matrices

$$\mathbf{T}_{Bal} = \mathbf{\Sigma}^{-1/2}\mathbf{U}^{\mathsf{T}}\mathbf{R}_Q \ , \ \mathbf{T}_{Bal}^{-1} = \mathbf{R}_P^{\mathsf{T}}\mathbf{V}\mathbf{\Sigma}^{-1/2}$$

B 4 : Calculate the similarity transformation (2)

The computation of $\mathbf{T}_{Bal}$ needs two matrix products and n scalar/vector operations only.

Starting point for improving this method is the use of the optimal scalar product [5] for the acurate computation of matrix products. A 400 decimal digit mantissa for the accumulator of the optimal scalar product requires only one rounding. Therefore the scalar product of

$$[1e55 \;\; 1 \;\; 1] \begin{bmatrix} -1 \\ 1 \\ 1e55 \end{bmatrix}$$

will be computed correctly and not essentially slower.

The computation of the Cholesky factors determines the accuracy of the whole method. Laub et al [7] proposed Hammarlings algorithm [3], which is a modification of the Bartels/Steward algorithm [1] for solving Lyapunov equations. The transformation on real Schur form is used and some sophisticated calculations lead to the Cholesky factors.

2 Improved Computation of $\mathbf{R}_P$ and $\mathbf{R}_Q$

A better way of calculating Cholesky factors without transforming the Lyapunov equation will now be shown. Starting point is the bilinear Lyapunov equation:

$$\mathbf{A}^{\mathsf{T}}\mathbf{X} + \mathbf{X}\mathbf{A} + \mathbf{W} = \mathbf{0}\,, \tag{11}$$

where $\mathbf{A}, \mathbf{W}, \mathbf{X} \in \mathbf{R}^{n \times n}$ $\mathbf{W}, \mathbf{X}$ symmetric and $\mathbf{X}$ pos. definite, therefore real Cholesky factors can be found:

$$\mathbf{X} = \mathbf{R}^{\mathsf{T}}\mathbf{R} \tag{12}$$

which change (11) to the nonlinear equation

$$\mathbf{A}^{\mathsf{T}}\mathbf{R}^{\mathsf{T}}\mathbf{R} + \mathbf{R}^{\mathsf{T}}\mathbf{R}\mathbf{A} + \mathbf{W} = \mathbf{0}\,, \tag{13}$$

The $n(n+1)/2$ unknown coeficients $r_{ij} \in \mathbf{R}$ are combined in a vector :

$$\mathbf{r}^{\mathsf{T}} = [r_{11}, r_{12}, \ldots, r_{1n}, r_{22}, \ldots, r_{2n}, \ldots, r_{nn}] = [r_1, \ldots, r_m, \ldots, r_{n(n+1)/2}]$$

The nonlinear equation (13) can be interpreted as a general vector function

$$\mathbf{f}(\mathbf{r}) = \mathbf{0} \tag{14}$$

The zeros of this nonlinear vector function are calculated with the well known Newton method, which requires the calculation of the so called Jacobi matrix:

$$\mathbf{J} = \begin{bmatrix} \frac{\partial f_1(\mathbf{r})}{\partial r_1} & \frac{\partial f_1(\mathbf{r})}{\partial r_2} & \cdots & \frac{\partial f_1(\mathbf{r})}{\partial r_{n(n+1)/2}} \\ \vdots & & & \vdots \\ \frac{\partial f_{ij}(\mathbf{r})}{\partial r_1} & & & \vdots \\ \vdots & & & \vdots \\ \frac{\partial f_{n(n+1)/2}(\mathbf{r})}{\partial r_1} & \cdots & \cdots & \frac{\partial f_{n(n+1)/2}(\mathbf{r})}{\partial r_{n(n+1)/2}} \end{bmatrix}, \tag{15}$$

The iterative Newton method

$$\mathbf{x}_{k+1} = \mathbf{x}_k - \mathbf{J}^{-1}(\mathbf{x}_k)\mathbf{f}(\mathbf{x}_k) \tag{16}$$

will converge if the starting vector is a sufficiently good approximation of the solution $\mathbf{f}(\mathbf{x}^*) = \mathbf{0}$. The inverse of $\mathbf{J}$ can be replaced by an approximation, see [8] for details.

$\mathbf{J}$ is calculated by partial differentiation of a general function:

$$f_{ij}(\mathbf{r}) = \mathbf{a}_i^{\mathsf{T}}\mathbf{x}_j + \mathbf{x}_i^{\mathsf{T}}\mathbf{a}_j + w(i,j) \qquad\qquad j \geq i \tag{17}$$

$\mathbf{x}$ can be replaced via:

$$\mathbf{X} = \begin{bmatrix} r_{11} & 0 & \cdots & 0 \\ r_{12} & r_{22} & \ddots & \vdots \\ \vdots & & \ddots & 0 \\ r_{1n} & \cdots & \cdots & r_{nn} \end{bmatrix} \begin{bmatrix} r_{11} & r_{12} & \cdots & r_{1n} \\ 0 & r_{22} & & \vdots \\ \vdots & \ddots & \ddots & \vdots \\ 0 & \cdots & 0 & r_{nn} \end{bmatrix} \tag{18}$$

$$\mathbf{X} = \begin{bmatrix} r_{11}\begin{bmatrix} r_{11} \\ \vdots \\ \vdots \\ r_{1n} \end{bmatrix}, & r_{12}\begin{bmatrix} r_{11} \\ \vdots \\ \vdots \\ r_{1n} \end{bmatrix} + r_{22}\begin{bmatrix} 0 \\ r_{22} \\ \vdots \\ r_{2n} \end{bmatrix}, & \cdots \end{bmatrix} \tag{19}$$

$$\mathbf{x}_i = r_{1i}\begin{bmatrix} r_{11} \\ \vdots \\ \vdots \\ r_{1i} \\ \vdots \\ r_{1n} \end{bmatrix} + r_{2i}\begin{bmatrix} 0 \\ r_{22} \\ \vdots \\ r_{2i} \\ \vdots \\ r_{2n} \end{bmatrix} + \ldots + r_{ii}\begin{bmatrix} 0 \\ \vdots \\ 0 \\ r_{ii} \\ \vdots \\ r_{in} \end{bmatrix} \tag{20}$$

$\mathbf{x}_i$ is a sum of vectors:

$$\mathbf{x}_i = r_{1i}\rho_1 + r_{2i}\rho_2 \ldots + r_{yi}\rho_y \ldots + r_{ii}\rho_i$$

The partial derivative of $\mathbf{x}_i$ is determined by the following scheme:

$$r_m \ni \mathbf{x}_i \qquad \frac{\partial \mathbf{x}_i}{\partial r_m} = 0 \tag{21}$$

$$r_m \in \mathbf{x}_i \wedge r_m \in \rho_y \qquad \frac{\partial \mathbf{x}_i}{\partial r_m} = \frac{r_{yi}\partial \rho_y}{\partial r_m} \tag{22}$$

$$r_m \in \mathbf{x}_i \wedge r_m \in \rho_y \wedge r_m = r_{yi} \qquad \frac{\partial \mathbf{x}_i}{\partial r_m} = \frac{\partial r_{yi}\rho_y}{\partial r_m} \tag{23}$$

The Jacobian is calculated with the optimal scalar product if (17) is formed as vector product:

$$f_m(\mathbf{r}) = \begin{bmatrix} \mathbf{a}_i^T & \mathbf{a}_j^T \end{bmatrix} \begin{bmatrix} \mathbf{x}_j \\ \mathbf{x}_i \end{bmatrix} + w_{ij}$$

where the partial derivative

$$\frac{\partial f_m(\mathbf{r})}{\partial r_m} = \begin{bmatrix} \mathbf{a}_i^T & \mathbf{a}_j^T \end{bmatrix} \frac{\partial \begin{bmatrix} \mathbf{x}_j \\ \mathbf{x}_i \end{bmatrix}}{\partial r_m} \tag{24}$$

needs $2(n+1)/2$ roundings only.

3 Numerical example

The improvement achieved via the proposed method is demonstrated with the Gramians of the 8 th order system:

$$G(s) = \frac{0.018s^6 + 0.0193s^5 + 0.637s^4 + 0.4s^3 + 4.2s^2 + 2.4s + 1}{0.000822s^8 + 0.00121s^7 + 0.0474s^6 + 0.0521s^5 + 0.841s^4 + 0.638s^3 + 4.31s^2 + 2.05s + 1}$$

$\mathbf{R}_P$ and $\mathbf{R}_Q$ are computed in four different ways:

W 1 : Compute the Gramians and calculate the Cholesky decomposition.

W 2 : Compute $\mathbf{R}_P$ and $\mathbf{R}_Q$ with the method of Hammarling.

W 3 : Compute $\mathbf{R}_P$ and $\mathbf{R}_Q$ with the new method.

W 4 : Calculate $\mathbf{R}_P$ and $\mathbf{R}_Q$ with the new method and optimal scalar product.

The programing language PC–MATLAB [11] is used for method W1, W3 and W2 is programmed with the NAG package. W4 is realized in the ACRITH FORTRAN subroutine package [4] on a SIEMENS mainframe. The results (see Table 1) show growing improvements from W1 to W4. Other calculated examples lead to the same order of accuracy. W4 is extendable by interval arithmetic to the computation of guaranteed results.

W 1	$\|\|f(\mathbf{r}_P)\|\|_1 = 2.28e-9$	$\|\|f(\mathbf{r}_Q)\|\|_1 = 2.63e-2$
W 2	$\|\|f(\mathbf{r}_P)\|\|_1 = 4.20e-10$	$\|\|f(\mathbf{r}_Q)\|\|_1 = 9.23e-6$
W 3	$\|\|f(\mathbf{r}_P)\|\|_1 = 6.77e-12$	$\|\|f(\mathbf{r}_Q)\|\|_1 = 3.81e-7$
W 4	$\|\|f(\mathbf{r}_P)\|\|_1 = 1.96e-13$	$\|\|f(\mathbf{r}_Q)\|\|_1 = 3.47e-8$

Table 1:

4 Conclusions

The procedure outlined in part 2 enables an improved computation of the Cholesky factors of the Gramians. The Cholesky factors are computed directly out of the Lyapunov equation without disadvantageous transforming the system. The new method can work with the original data $(\mathbf{A}, \mathbf{B}, \mathbf{C})$ and calculate $\mathbf{R}_P$ and $\mathbf{R}_Q$ with high accuracy.

The computation of the Cholesky factors determines the accuracy of the balancing algorithm at all. The new method and the use of the optimal scalar product will extend the application of balanced realizations as a model reduction procedure for minimal systems.

References

[1] R.H. Bartels and G.W. Stewart. Solution of the matrix equation $AX + BX = C$. *Commun. Ass. Comput. Mach.*, 15:820–826, 1972.

[2] K. Glover. All optimal Hankel-norm approximations of linear multivariable systems and their L^∞ error bounds. *Int. J. Contr.*, 39(6):1115–1193, 1984.

[3] S. J. Hammarling. Numerical solution of the stable, non–negative definite lyapunov equation. *IMA J. Numer. Anal.*, 2:303–323, 1982.

[4] IBM. *High–accuracy arithmetic subroutine library.* April 1986.

[5] U.W. Kulisch and W.L. Miranker. *A new approach to scientific computation.* Academic Press, New York, 1983.

[6] A.J. Laub. On computing balancing transformations. In *1980 Joint Automat. Contr. Conf,* San Francisco, CA, August 1980. Session FA8-E.

[7] A.J. Laub, M.T. Heath, C.C. Paige, and R.C. Ward. Computation of system balancing transformations and other applications of simultaneous diagonalization algorithms. *IEEE Trans. Autom. Control AC-32*, 115–121, 1987.

[8] G. Ludyk. *CAE von dynamischen Systemen.* Springer — Verlag Berlin, 1989.

[9] B.C. Moore. Principal component analysis in linear systems: controlability, observability, and model reduction. *IEEE Trans. Autom. Control AC-26*, 17–32, 1981.

[10] M.G. Safonov and R.Y. Chiang. A Schur method for balanced – truncation modell reduction. *IEEE Trans. Autom. Control AC-34*, 729–733, 1989.

[11] Inc. The MathWorks. *PC - MATLAB and CONTROL SYSTEM TOOLBOX USER'S GUIDE. Version 3.2.* June 1987.

Adaptive Control of Linear Plants With Unknown High-Frequency Gain

Jürgen Cordes

Abstract

Based on a control concept by [5] it is tried to use this controller in practice. Unknown first order plants are simulated to check the behaviour of the regulator in perturbed surrounding. The concept is then used as a set point controller. It is shown that in practical use this adaptive control concept failed.

1 Introduction

This paper is based on a first order plant with unknown parameters a and b :

$$\dot{x}(t) = ax(t) + bu(t) \quad \text{mit } x(0) = x_0 \quad \text{und } b \neq 0 \tag{1}$$

For such processes the literature gives four common assumptions which can be used as sufficient conditions for global stability in every adaptive control scheme (state of 1984):

I.) The zeros of the system to be controlled must be in the left open half of the complex plane.

II.) The relative degree of the system must be known.

III.) An upper bound for the order of the system must be known.

IV.) The sign of the high-frequency gain must be known.

Concerning No.IV of the above list Morse [1] assumed in 1983 that with two functions f and g

$$f : \mathbf{R}^{m+1} \to \mathbf{R} \;\; k = f(x,k) \quad ; \quad g : \mathbf{R}^{m+1} \to \mathbf{R} \;\; u = g(x,k) \tag{2}$$

a system could <u>not</u> be stabilized, if the controller has no information about the sign of b. For a special case he proved his conjecture.

Nussbaum [2] gave a counterexample to the conjecture of Morse with the following two functions

$$f(x,k) = x\left(k^2+1\right) \qquad g(x,k) = x\,(x^2+1)\,h(k)$$
$$\text{with } h(k) = \cos(\tfrac{\pi}{2}k)\,exp\,(k^2) \tag{3}$$

He proved that $x(t)$ and $k(t)$ are bounded for $t \to \infty$.

This concept had been extended by Morse [3] for systems of higher order with the relative degree one or two. He used therefore

$$f(x,k) = x^2 \quad ; \quad g(x,k) = xk^2\cos(k) \tag{4}$$

Willems and Byrnes [4] used the Nussbaum concept with the oscillating gain for systems of arbitrary and unknown order but with relative degree one. Additionally all zeros must be in the open left half plane. They used :

$$h(k) = \begin{cases} +1 & \text{for } n^2 \le k \le (n+1)^2 \quad \text{for } n = 0,2,4,\ldots \\ -1 & \text{for } n^2 \le k \le (n+1)^2 \quad \text{for } n = 1,3,5,\ldots \end{cases} \tag{5}$$

Summarized there are three disadvantages for these control concepts:

1.) The dynamic behaviour of the controlled plant is not predictable as the controller converges.

2.) The controller gain tends to infinity as the initial perturbation of the system is increased.

3.) Stable processes will be periodically destabilized and will behave with a completely different dynamic.

2 The concept by Heymann et.al.

The control concept by Heymann et.al. [5] has the following structure (Figure 1):

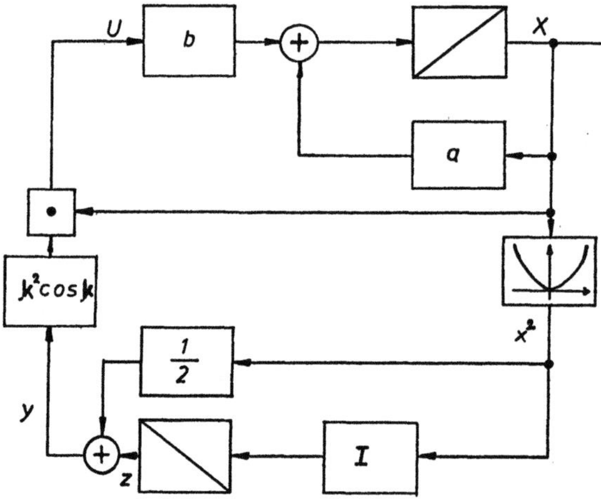

Figure 1: The control structure and the process

$$\dot{z} = \quad Ix^2 \quad \text{with } I > 0 \tag{6}$$

$$k = \quad \xi(x,z) \quad \text{with } \xi = \frac{x^2}{2} + z \tag{7}$$

$$u = \quad xg(k) \quad \text{with } g(k) = k^2\cos(k) \tag{8}$$

If the Willems-Byrnes control scheme is used with the above switching function (8) then the main difference between both controllers is the proportional part $\frac{x^2}{2}$ in equation (7).

Discovering the dynamics of the plant with the controller leads to two coupled differential equations:

$$\dot{x} = \left(a + bk^2 \cos(k)\right) x \tag{9}$$

$$\dot{k} = \left(I + a + bk^2 \cos(k)\right) x^2 \tag{10}$$

It is easy to see that the only equilibrium point of the closed loop system is $x = 0$.

Theorem 1: For the plant made from (9) and (10) with arbitrary x_0 and k_0 the following can be shown:

a) $x(t)$ ist bounded on $t \in [t_0, \infty)$

b) $\lim_{t \to \infty} x(t) = 0$

Proof : Integration of 10 gives:

$$k(t) = k_0 - \frac{x_0^2}{2} + \frac{x^2}{2} + \int_{t_0}^{t} I x^2(\tau)\, d\tau$$

But if $k(t)$ is bounded, which is shown below, then every part of the sum above must stay bounded. This holds also for the integral. The integrand $x^2(t)$ reaches zero.

■

Out of $x(t) \neq 0$ the gain $k(t)$ can reach an equilibrium point in every solution of

$$I + a + bk^2 \cos(k) = 0 \Leftrightarrow g(k) = k^2 \cos(k) = -\frac{a + I}{b} \tag{11}$$

But actually only every second solution is an equilibrium point for $k(t)$.

We can see in figure 2 both parts of (11), the switching function and the function of a straight line. At the crossing points of both functions we find the equilibrium points. If we additionally examine the signs of the first derivative of $k(t)$ we can see that a special $k(t)$, e.g. the point I in figure 2, tends to fixed $k(\infty)$. So every equilibrium point has its left and right side area of attraction. The point inbetween with $\dot{k} = 0$ will not be met in practical use. Caused by the periodicity of $g(k)$ there will be for every $k(0)$ a bounded $k(\infty)$ (Proof in [5]).

The phase plane (fig.3) shows this behaviour for different initial conditions x_0. At a fixed time $k(t)$ stays constant. If the controller compensates the initial condition with $-\frac{x_0^2}{2}$ then every x_0 leads to one fixed gain $k(\infty)$ (fig.4). The differences between these two concepts will be shown if we look at the behaviour in time for an initial condition. If we look at the velocity of the decaying $x(t)$ we can see that the non-compensated system reaches the equilibrium faster than the other. The initial gain leads to a faster behaviour. The compensated system passes a short period of destabilization before it converges on a lower level for the gain $k(t)$. So the controller with the compensation needs more time to converge (fig. 5 and fig.6).

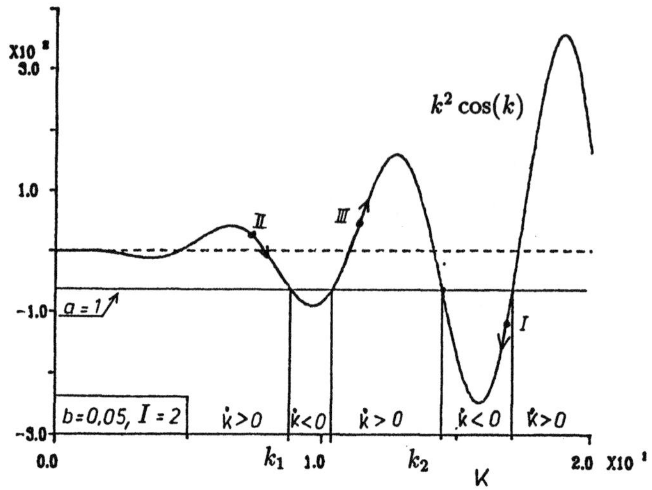

Figure 2: Switching function and equilibrium points

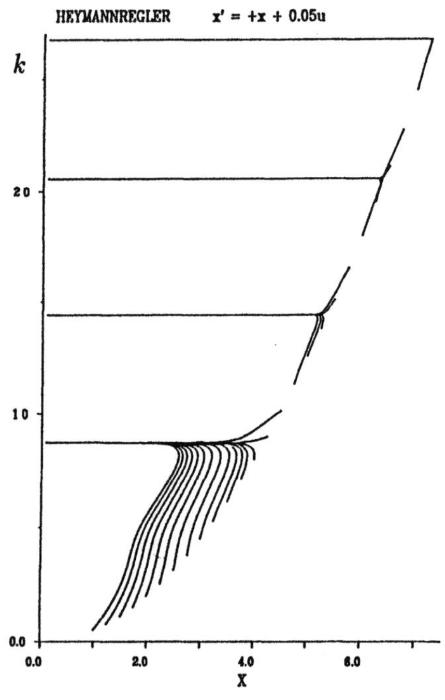

Figure 3: Phase plane without compensation

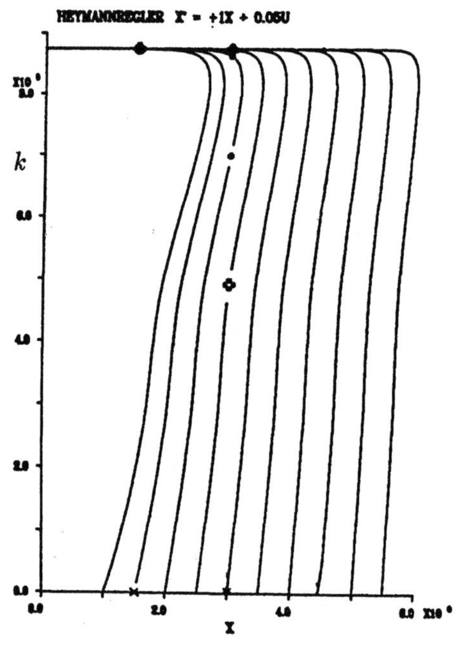

Figure 4: Phase plane with compensation

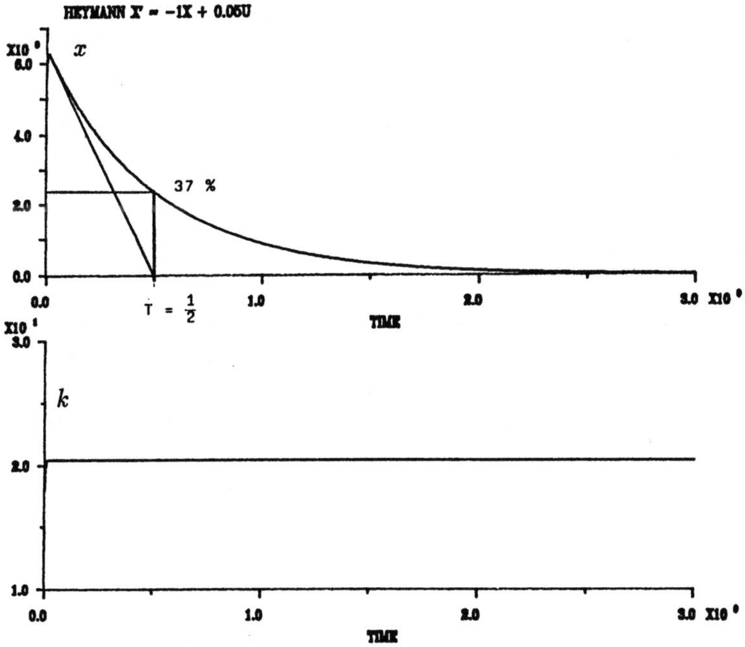

Figure 5: Simulation of a stable system with $k_0 \neq 0$

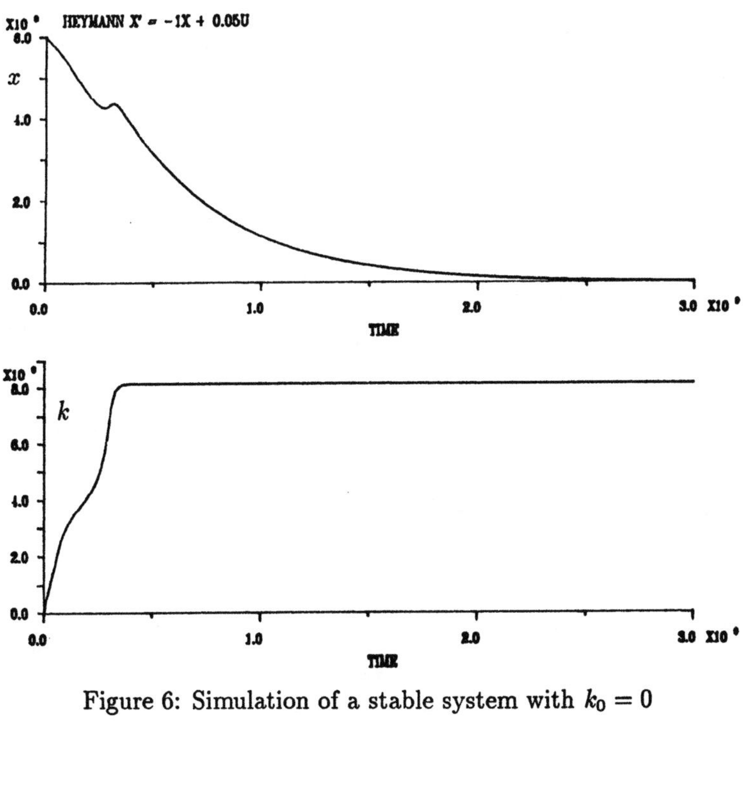

Figure 6: Simulation of a stable system with $k_0 = 0$

If we look at the equilibrium point of the controller while the plant is still in motion we can see an other positive aspect of this control concept. From eq.7 we get by derivation:

$$\dot{k} = x\dot{x} + Ix^2 \rightarrow \dot{x} = -Ix \quad \text{with } \dot{k} = 0 \quad \text{and } x \neq 0 \tag{12}$$

After the convergence of the controller we can specify the dynamic of the closed-loop by choosing I. Fig.5 shows a simulation with $I = 2$ which gives a time constant of $T = 0.5$.

3 Perturbations

Investigating the possibility for practical application of this new control concept we have to look at its behaviour in a noisy surrounding caused by e.g. measurement noise. If we simulate with additive noise at the output of the plant we get a very negative result. The simulation and the theoretic investigation as well show that the closed-loop will be destabilized.

Integrating eq.12 gives

$$\begin{aligned} k &= k_0 - \tfrac{x_0^2}{2} + \tfrac{x_a^2}{2} + \int_{t_0}^{t} I x_a^2(\tau)\, d\tau \\ &\quad \text{with } x_a(t) = x(t) + R(t) \\ &\quad R(t) = \text{perturbation function} \end{aligned} \tag{13}$$

Because $x_a^2(t) > 0$ and unequal to zero the integral is increasing and so $k(t)$ will <u>not</u> stay bounded. The Simulation (fig. 7) shows that the increase of k(t) is responsible for the periodic behaviour of $x(t)$.

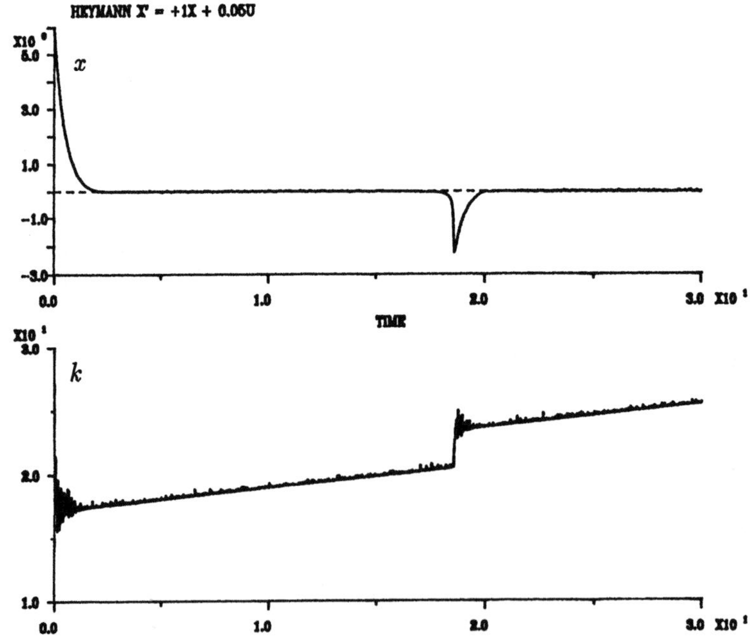

Figure 7: Simulation of an unstable system with perturbations

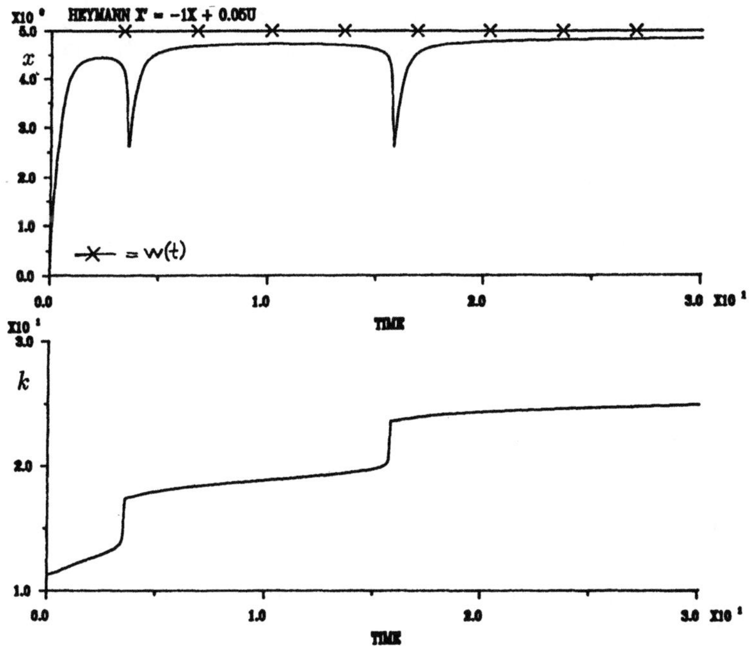

Figure 8: Simulation of a stable system with set-point control

4 Set-Point Control

The structure of the plant with the controller is shown in fig.9. The input for the controller is now the control-difference. For the following investigation we use $w(t) = \sigma(t)$.

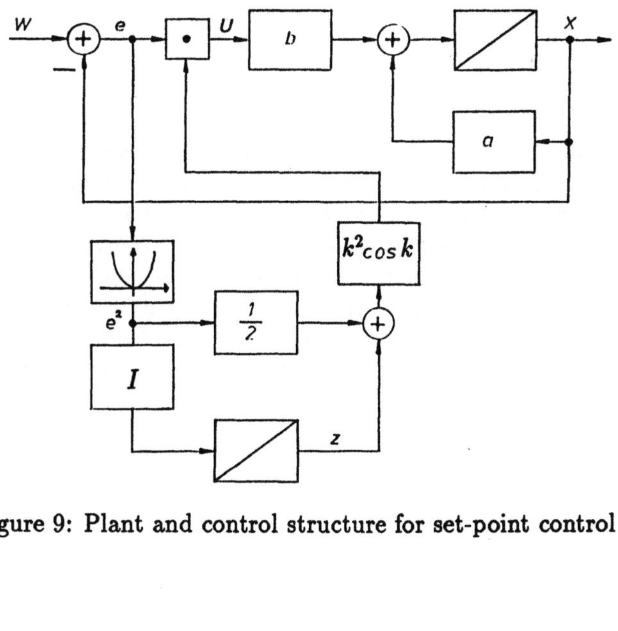

Figure 9: Plant and control structure for set-point control

Theorem 2: For the plant shown in fig.9 with $(a, b) \in (0,0), x_0$ and k_0 arbitrary, $I \in \mathbf{R}^+\backslash\{0\}$ and $w(t) = \sigma(t)$ unit jump-function in t_0:
Then we get:

a) $\sup_{t>t_0} k(t) = +\infty$

b) $\inf_{t>t_0} k(t) = -\infty$

Proof :

1.) $\dot{x}(t) = ax(t) + bg(k)(w(t) - x(t))$

2.) Target of the set-point control:
$\lim_{t\to\infty} (w(t) - x(t)) \to 0$

3.) $| k(t) |$ is bounded

From (2) we get in (1): (because w =constant and $a = 0$)
$\dot{x} \to aw$ for $t \to \infty$
So $x(t) \to \infty$ for $t \to \infty$
But this is not the intention of set-point control. (2) could only hold with:
$| k(t) | \to \infty$ and therefore $| g(k) | \to \infty$

■

The simulation in fig.8 demonstrates this behaviour. $k(t)$ increases unbounded and the state $x(t)$ is periodically destabilized. We can also get from this simulation that this adaptive control concept works like a conventional proportional controller with adjustable gain. The increasing gain reduces control difference $e(t)$ which can be determined by the decreasing distance between $w(t)$ and $x(t)$.

5 Conclusions

We can get two different kinds of results from the investigation of the adaptive control concept by [5]. On one side we have great theoretical advantages, e.g. bounded gain, constant dynamic after adaptation and good dynamic behaviour before the adaptation. The other side shows us the disadvantages of being very sensitve against perturbations. Very small noise at the output of the plant prevents an equilibrium of the controller. The use as a set-point controller, which is a frequently used application in practice, is impossible because the controller does not reach an equilibrium point. Some changes in the basic idea of this adaptive control concept must show in future whether this kind of control has a chance in practical application.

References

[1] A.S. Morse : *Recent Problems in Parameter adaptive Control*
I.D. Landau Editions du CNRS Vol.: 3 1983

[2] R.D. Nussbaum : *Some Remarks on a Conjecture in Parameter adaptive Control*
Systems & Control Letters Vol.:3 1983

[3] A.S. Morse : *An adaptive Control for globally stabilizing linear systems with unknown high-frequency Gain*
Report No.: 8402 Department of electrical Engineering Yale University New Haven 1983

[4] J.C. Willems, C.I. Byrnes : *Global adaptive Stabilization in the Absence of Information on the Sign of the high-frequency Gain*
Lecture Notes in Control & Information Sciences Vol.:62 Part1 1984

[5] M. Heymann, J.H. Lewis, M. Field : *Remarks on the adaptive Control of linear Plants with unknown high-frequency Gain*
Systems & Control Letters No.:5 1985

[6] J. Cordes, H.-W. Philippsen : *Adaptive Regelung linearer Prozesse mit unbekanntem high-frequency gain*
Diplomarbeit University of Bremen 1986

Design of a Full-Order Observer and its Minimal-Order Version for a One-Link Flexible Robot Arm

C. Bruce-Boye

Abstract

A full-order observer and its minimal-order version are designed with reference to an approximate finite-dimensional linear, time-invariant model of a single-link flexible robot arm. By use of the principle of duality, numerically reliable algorithms are applied in order to compute the observer gain matrix. Since several linearly independent plant outputs are available for measurement, the observer gain matrix is not unique. The resulting degrees of freedom are used to minimize the observer gain matrix, thereby improving the behaviour of the closed loop system incorporating the state observer. Simulation tests to validate the effectiveness of the observers in the closed loop system are included.

1. Introduction:

One aim in the design of modern manipulators for industrial applications, and especially for operations in space, is a light structure design, which usually results in high-speed performance and low energy consumption. Another consequence is the oscillatory behaviour of these flexible structures with respect to the payload. Nevertheless the controlled motion of the robot arm must be accurate and without vibrations. Therefore a more sophisticated control law must be implemented which takes into account not only the rigid modes but also the vibration modes. Our goal is to compensate the oscillatory behaviour of the flexible one-link robot arm by linear state feedback. This implies the need for a state observer [7]. A state observer is an additional dynamical system which is driven by the input and, in general, by the output of the original system. The purpose of the observer is to reconstruct all states of the original system where those states are difficult to measure directly. In order to control the flexible modes of the robot arm, a complete model of the flexible structures is needed. The dimension of such a model is infinite [1]. For reasons of physical implementation, we finally deal with a tenth-order, reduced-order model [1,2,3].

2. Problem description:

Consider the linearized dynamical plant of the flexible robot arm which is modelled by state, output, and control-law equations of the respective forms,

$$\dot{\mathbf{x}}(t) = \mathbf{A}\mathbf{x}(t) + \mathbf{B}\mathbf{u}(t) \quad (1)$$

$$\mathbf{y}(t) = \mathbf{C}\mathbf{x}(t) \quad (2)$$

$$\mathbf{u}(t) = -\mathbf{K}\mathbf{x}(t) \quad (3)$$

where $\mathbf{x}$ is the n-dimensional state vector, $\mathbf{u}$ is the p-dimensional input vector and $\mathbf{y}$ is the m-dimensional output vector of the plant. All matrices are of appropriate dimensions and the measure matrix $\mathbf{C}$ is of full rank m.

Furthermore, the matrix pair $(\mathbf{A}, \mathbf{C})$ is completely observable [3] and the matrix pair $(\mathbf{A}, \mathbf{B})$ is completely controllable [10].

The order ten of the system is relatively high. According to [15], the system matrix is ill-conditioned. Thus one problem that arises is to find computationally stable methods in order to compute the required observer gain matrix. In our case, when an infinite-dimensional plant has to be controlled by a finite-dimensional controller, care must be taken not to stimulate unmodelled flexible modes by the observer or by the contoller itself. Therefore a safety margin must be kept between the lowest frequency of the unmodelled modes and the highest frequency of the modelled modes. This can be achieved by a suitable pole placement [1]. A numerically reliable technique is offered by the Riccati equation. But then the location of the poles can be influenced only indirectly by the weighting matrices. If the principle of duality is applied , numerically reliable pole placement algorithms are available. We chose three of them: the robust pole assignment "place" by Kautzky and Nichols [6], the modal approach "porter" by Porter and Crossly [5], and the modal synthesis "vomosy" by Roppenecker [4]. The choice of the pole locations provides some degree of freedom to the designer. However, because of the safety margin mentioned before, this choice is additionally limited. That is why we favor the modal synthesis by Roppenecker. His method explicitly offers all additional degrees of freedom, apart from that of the pole locations, by the so-called invariant parameter vectors. Fortunately a number of different outputs of the plant can be measured. We use the resulting different solutions to minimize the norm $\|\cdot\|$ [1] of the observer gain matrix. The behaviour of the closed loop system incorporating the state observer was thus improved.

3. State observer:

A dynamical system of dimension ϑ that reconstructs the plant state vector $\mathbf{x(t)}$ in (1), (2) in real time, is called a state observer. It can be expressed by the following equations

$$\dot{\mathbf{z}}(t) = \mathbf{Dz}(t) + \mathbf{Mu}(t) + \mathbf{Ly}(t) \tag{4}$$
$$\hat{\mathbf{x}}(t) = \mathbf{S}_1\mathbf{z}(t) + \mathbf{S}_2\mathbf{y}(t)\,, \tag{5}$$

where $n \geq \vartheta \geq n - m$ and $\mathbf{z}(t)$ is the ϑ-dimensional state vector of the observer with a (ϑ, ϑ)-dimensional matrix $\mathbf{D}$. $\mathbf{M}, \mathbf{L}$ are matrices of the dimensions (ϑ, p) and (ϑ, q) respectively. They are weighting matrices of the measured input and output values of the plant. The required reconstructed state vector $\hat{\mathbf{x}}(t)$ is a (n, ϑ)-dimensional linear combination $\mathbf{S}_1$ of the observer states and a (n, q)-dimensional linear combination $\mathbf{S}_2$ of the output $\mathbf{y}(t)$ of the plant. A successful reconstruction of the plant states is given if the five well known conditions are fulfilled.

[1] $\|\cdot\| = \sqrt{\mathbf{x}^T\mathbf{x}}$ for vectors and $\left(\sum_{i=1}^{m}\sum_{j=1}^{n}|a_{ij}|^2\right)^{\frac{1}{2}}$ for matrices

$B_1 \qquad n \geq \vartheta \geq n - m$

$B_2 \qquad \mathbf{DT} - \mathbf{TA} = -\mathbf{LC}$

$B_3 \qquad \mathbf{S}_1\mathbf{T} + \mathbf{S}_2\mathbf{C} = \mathbf{I}_n \Longleftrightarrow \quad \mathrm{rg}\begin{pmatrix} \mathbf{T} \\ \mathbf{C} \end{pmatrix} = \mathrm{n}$

$B_4 \qquad \mathbf{M} = \mathbf{TB}$

$B_5 \qquad \Re e\, \lambda_i(\mathbf{D}) < 0 \qquad i = 1, \ldots, n\ .$

A successful reconstruction means that the observer output $\hat{\mathbf{x}}(t)$ converges to the original state vector $\mathbf{x}(t)$ for any input $\mathbf{u}(t)$ that is

$$\lim_{t \to \infty} \mathbf{e}(t) = \mathbf{0} \tag{6}$$

where

$$\mathbf{e}(t) = \mathbf{x}(t) - \hat{\mathbf{x}}(t) \tag{7}$$

is the reconstruction error which satisfies the homogeneous differential equation

$$\dot{\mathbf{e}}(t) = \mathbf{De}(t); \quad \mathbf{e}(0) = \mathbf{e}_0\ . \tag{8}$$

3.1 Identity observer

If we choose the order ϑ of the observer to be equal to the order n of the plant model, the tranformation matrix $\mathbf{T}$ will represent the identity matrix $\mathbf{I}_n$. Thus the conditions B_2 to B_4 lead to the following description of the identity observer.

$$\dot{\hat{\mathbf{x}}}(t) = \mathbf{D}\hat{\mathbf{x}}(t) + \mathbf{Bu}(t) + \mathbf{Ly}(t); \qquad \hat{\mathbf{x}}(t_0) \stackrel{!}{=} \mathbf{0} \tag{9}$$

Now the observer gain matrix $\mathbf{L}$ has to be designed in such a way that all eigenvalues of the coefficient matrix $\mathbf{D}$ are in the left half-plane. This is always possible if the matrix pair $(\mathbf{A}, \mathbf{C})$ is completely observable.

3.2 State observer in the closed loop system

In the closed loop system where the original state vector $\mathbf{x}(t)$ is subtituted by the reconstucted state $\hat{\mathbf{x}}(t)$ in (3), we obtain the new control-law.

$$\hat{\mathbf{u}}(t) = -\mathbf{K}\hat{\mathbf{x}}(t) \tag{10}$$

If the equation above is split up into

$$\hat{\mathbf{u}}(t) = \mathbf{u}(t) + \mathbf{u}_{error}(t)\ ,$$

it follows from (7) that

$$\mathbf{u}_{error}(t) = \mathbf{Ke}(t) \tag{11}$$

is the control error. This error exists only during the converging phase of the observer. It can cause the closed loop system to behave in a strange manner. This can be avoided by limiting the control error. We obtain the inequality

$$\|\mathbf{u}_{error}(t)\| \leq \|\mathbf{K}\|\|\mathbf{e}(\mathbf{t})\| \qquad \text{for all } t \geq t_0 \tag{12}$$

directly upon taking norms of (11). Now it is obvious that the control error is minimized if

$$\|\mathbf{K}\| \quad \rightarrow \quad min \tag{13}$$

and / or

$$\|\mathbf{e}(t)\| \quad \rightarrow \quad min \qquad \text{for all } t \geq t_0\,. \tag{14}$$

We assume that all eigenvalues of the closed loop system are given and that the system (1), (2) in question is a single input and multiple output system. Thus we concentrate on (14).

4. Limiting the reconstruction error

The solution of (8) can be expressed as

$$\mathbf{e}(t) \;=\; exp[(\mathbf{A}-\mathbf{LC})t)]\mathbf{e}(0); \quad \mathbf{e}(0)=\mathbf{x}(0) \tag{15}$$

Then the error norm $\|\mathbf{e}(t)\|$ satisfies

$$\begin{aligned}
\|\mathbf{e}(t)\| \;&=\; \|exp[(\mathbf{A}-\mathbf{LC})t)]\mathbf{e}(0)\|; \quad \|\mathbf{e}(0)\| = \|\mathbf{x}(0)\| \\
&\leq\; \|exp[(\mathbf{A}-\mathbf{LC})t)]\|\;\|\mathbf{e}(0)\| \\
&\leq\; exp[(\|\mathbf{A}\|+\|\mathbf{LC}\|)t)]\;\|\mathbf{e}(0)\|
\end{aligned}$$

and thus

$$\|\mathbf{e}(t)\| \;\leq\; exp[(\|\mathbf{A}\|+\|\mathbf{L}\|\|\mathbf{C}\|)t)]\;\|\mathbf{e}(0)\| \tag{16}$$

This demonstrates that minimizing the norm of the observer gain matrix $\|\mathbf{L}\|$ also minimizes a bound on the reconstruction error $\|\mathbf{e}(\mathbf{t})\|$ for all t in the time interval $[0, t_1]$, where t_1 is the shortest time after the error has decayed.

5. Pole placement algorithms

The duality between the homogeneous equations

$$\dot{\mathbf{z}}(t) = (\mathbf{A}-\mathbf{LC})\mathbf{z}(t)$$

and

$$\dot{\bar{\mathbf{z}}}(t) = (\mathbf{A}^T - \mathbf{C}^T\mathbf{L}^T)\bar{\mathbf{z}}(t)\,,$$

leads to the use of pole assignment algorithms for the observer design. Then $\mathbf{A} \equiv \mathbf{A}^T$, $\mathbf{B} \equiv \mathbf{C}^T$ and the solution $\mathbf{L} \equiv \mathbf{K}^T$.

The pole placement algorithm by Kautzky and Nichols computes a robust solution to the multi-input state feedback pole assignment problem. The observer gain matrix is obtained by assigning n linearly independent eigenvectors corresponding to the required eigenvalues (poles), so that the matrix of the eigenvectors $\mathbf{V}_{(n,n)}$ is as well-conditioned as possible [6,15].

$$\begin{aligned}\kappa_2(\mathbf{V}) &= \|\mathbf{V}\|_2\|\mathbf{V}^{-1}\|_2 \\ \kappa_2(\mathbf{V}) &\rightarrow 1\end{aligned}$$

It is shown in [6] that the assigned poles are then as insensitive to perturbations as possible (robust) and the resulting feedback matrix is reasonably bounded.

The modal synthesis by Roppenecker [4] offers all degrees of freedom to the designer by means of the q-dimensional parameters vectors.

$$\mathbf{p}_i^T = \mathbf{v}_{Di}^T\mathbf{L} \qquad i = 1,\ldots,n \tag{17}$$

To simplify the presentation, let us assume that all eigenvalues $\lambda_{Di},\ldots,\lambda_{Dn}$ of $(\mathbf{A}-\mathbf{LC})$ are distinct and different from the eigenvalues of the plant. Then n linearly independent left eigenvectors $\mathbf{v}_{Di},\ldots,\mathbf{v}_{Dn}$ of the observer satisfying the equations

$$\mathbf{v}_{Di}^T(\mathbf{A}-\mathbf{LC}-\lambda_{Di}\mathbf{I}_n) = \mathbf{0}^T \qquad i = 1,\ldots,n \tag{18}$$

$$\mathbf{v}_{Di}^T(\mathbf{A}-\lambda_{Di}\mathbf{I}_n) = \mathbf{v}_{Di}^T\mathbf{LC} \qquad i = 1,\ldots,n \tag{19}$$

can be found. Together with (17), equation (19) becomes

$$\mathbf{v}_{Di}^T(\mathbf{A}-\lambda_{Di}\mathbf{I}_n) = \mathbf{p}_i^T\mathbf{C} \qquad i = 1,\ldots,n\,.$$

By a suitable choice of the observer eigenvalues we produce the observer eigenvectors as

$$\mathbf{v}_{Di}^T = \mathbf{p}_i^T\mathbf{C}(\mathbf{A}-\lambda_{Di}\mathbf{I}_n)^{-1} \qquad i = 1,\ldots,n\,. \tag{20}$$

Equation (17) is equivalent to

$$\mathbf{P} = \mathbf{V}_D\mathbf{L}\,. \tag{21}$$

If the parameter vectors are chosen in such a way that n linearly independent left eigenvectors are generated then the observer gain matrix results from

$$\mathbf{L} = \mathbf{V}_D^{-1}\mathbf{P}\,, \tag{22}$$

or in detail

$$\mathbf{L} = \begin{pmatrix} \mathbf{p}_1^T\mathbf{C}(\mathbf{A}-\lambda_{D1}\mathbf{I}_n)^{-1} \\ \vdots \\ \mathbf{p}_n^T\mathbf{C}(\mathbf{A}-\lambda_{Dn}\mathbf{I}_n)^{-1} \end{pmatrix}^{-1} \begin{pmatrix} \mathbf{p}_1^T \\ \vdots \\ \mathbf{p}_n^T \end{pmatrix}. \tag{23}$$

The observer gain matrix is now parameterized by the observer eigenvalues and the n, q-dimensional parameter vectors.

$$\mathbf{L} = \mathbf{L}(\lambda_{D1} \dots \lambda_{Dn}, \mathbf{p}_1 \dots \mathbf{p}_n) \tag{24}$$

For the general case see Roppenecker [4]. That solution differs mainly in two aspects from the modal approach by Porter und Crossly [5]. Firstly, there modal synthesis is based on the eigenvectors of the matrix $\mathbf{A}$, as distinguished from the method described above, where the eigenvectors of $(\mathbf{A} - \mathbf{LC})$ are required. Secondly, there is only one parameter vector of the dimension q left for the designer to choose.

Now the norm of the observer gain matrix can be minimized by "vomosy" [9]. Since the choice of the observer eigenvalues is limited with respect to the unmodelled flexible modes, we assume them to be given. Thus the performance index

$$J_n = \frac{1}{2} \sum_{i=1}^{n} \sum_{j=1}^{q} g_{ij} l_{ij}^2 \qquad g_{ij} \geq 0$$

is a function of the parameter vectors only. The numerical minimization of J_n by a gradient based search technique then yields the observer gain matrix with mimimal norm.

6. Numerical example

The linearized dynamical plant of the one-link flexible robot arm in the continuous-time case for a payload of 10 kg is given by the following model, where the system matrix $\mathbf{A}$ [1] is:

$$\left(\begin{matrix} 0 & 0 & 0 & 0 & 0 \\ 0 & 0 & 0 & 0 & 0 \\ 0 & 0 & 0 & 0 & 0 \\ 0 & 0 & 0 & 0 & 0 \\ 0 & 0 & 0 & 0 & 0 \\ 0 & 2.5743E+3 & 2.2317E+4 & 6.0147E+4 & 2.1773E+4 \\ 0 & -2.3722E+3 & -1.1383E+4 & -2.8908E+5 & 1.7921E+4 \\ 0 & -1.6776E+3 & -2.4503E+3 & -4.2767E+5 & 3.1517E+4 \\ 0 & -7.7031E+2 & -2.2630E+3 & -1.9629E+5 & 7.3432E+3 \\ 0 & 1.5421E+3 & -2.1974E+4 & 4.5682E+5 & -6.9426E+4 \end{matrix}\right.$$

$$\left.\begin{matrix} 1 & 0 & 0 & 0 & 0 \\ 0 & 1 & 0 & 0 & 0 \\ 0 & 0 & 1 & 0 & 0 \\ 0 & 0 & 0 & 1 & 0 \\ 0 & 0 & 0 & 0 & 1 \\ 0 & 1.8020E+1 & 1.5622E+2 & 4.2103E+2 & 1.5241E+2 \\ 0 & -1.6606E+1 & -7.9681E+1 & -2.0235E+3 & 1.2545E+2 \\ 0 & -1.1743E+1 & -1.7152E+1 & -2.9937E+3 & 2.2062E+2 \\ 0 & -5.3921E+0 & -1.5841E+1 & -1.3740E+3 & 5.1402E+1 \\ 0 & 1.0795E+1 & -1.5382E+2 & 3.1977E+3 & -4.8598E+2 \end{matrix}\right) \tag{25}$$

The state vector contains the rigid body coordinate $\varphi(t)$ in degrees, the time-dependent parts $\mathbf{v}(\mathbf{t})$ in 1/meter of the flexible coordinates $\mathbf{v}(\mathbf{x}, \mathbf{t})$, and the derivative with respect to the time of both coordinates. [1], [2].

$$\mathbf{X}^T = (\ \varphi \quad v_1 \quad v_2 \quad v_3 \quad v_4 \quad \dot{\varphi} \quad \dot{v}_1 \quad \dot{v}_2 \quad \dot{v}_3 \quad \dot{v}_4\ ) \tag{26}$$

The input vector is given as:

$$\mathbf{b} = \begin{pmatrix} 0 \\ 0 \\ 0 \\ 0 \\ 0 \\ 1.4066E+1 \\ -1.0023E+1 \\ -2.6028E+0 \\ -4.4626E-1 \\ 1.0674E+0 \end{pmatrix}. \tag{27}$$

The first row of the output matrix contains the second derivative of the shape function $\mathbf{v}"(\mathbf{x_M})$ of the flexible coordinates $\mathbf{v}(\mathbf{x},\mathbf{t})$ with respect to the location where the bending is measured. [1], [2].

$$\mathbf{C} = \begin{pmatrix} 0 & 4.4571E+0 & 2.3307E+1 & 3.2820E+1 & 2.4028E+1 & 0 & 0 & 0 & 0 & 0 \\ 1 & 0 & 0 & 0 & 0 & 0 & 0 & 0 & 0 & 0 \\ 0 & 0 & 0 & 0 & 0 & 1 & 0 & 0 & 0 & 0 \end{pmatrix} \tag{28}$$

We accept the poles of the closed loop system as suggested by [1]. Those eigenvalue locations yield a control matrix:

$$\mathbf{k}^T = \big(2.59E-2 \quad 1.82E+3 \quad -1.62E+3 \quad 4.79E+5 \quad -3.07E+4$$
$$3.36E-1 \quad 1.23E+1 \quad -1.52E+1 \quad 3.23E+3 \quad -2.16E+2 \big). \tag{29}$$

We place the eigenvalues of the observer to:

eigenvalue			
plant		observer	
Re	Im	Re	Im
0.0	0.0	-4.5628E+0	0.0
0.0	0.0	-4.5628E+0	0.0
-1.3979E+3	0.0	-1.3979E+3	0.0
-1.4169E+2	±1.4285E+2	-1.4169E+2	±1.4285E+2
-2.2849E+1	±7.7500E+1	-2.2849E+1	±7.7500E+1
-3.8253E+0	±3.2837E+1	-1.8950E+1	±4.2510E+1
-1.5912E+2	0.0	-1.5912E+2	0.0

Table 1: Plant and observer eigenvalues, respectively

The initial condition error $\mathbf{e}_0 = \mathbf{x}_0$ is shown in table 2

-9.00E+1	degree
-1.04E-1	1/m
-4.42E-3	1/m
5.22E-4	1/m
2.81E-3	1/m
1.81E+0	degree/sec
-5.06E-2	1/m sec
7.49E-4	1/m sec
1.85E-4	1/m sec
-1.41E-3	1/m sec

Table 2: Initial condition error

The following observer gain matrices are found by the algorithms.

"place":

$$\mathbf{L} = \begin{pmatrix} 3.41E-1 & 2.02E+1 & 7.05E-1 \\ 7.07E+0 & 6.15E+1 & -1.84E+0 \\ 7.90E-1 & -1.44E+1 & -3.51E-1 \\ 1.54E-2 & -2.30E+0 & -1.59E-1 \\ 8.09E-1 & 5.60E+1 & -1.37E-1 \\ 1.45E+3 & -2.02E+3 & -5.07E+1 \\ -4.12E+3 & 8.15E+3 & 1.84E+2 \\ -5.82E+3 & 1.03E+4 & 3.65E+2 \\ -2.75E+3 & 4.31E+3 & 2.08E+2 \\ 5.61E+3 & -1.11E+4 & -2.91E+2 \end{pmatrix}. \tag{30}$$

"porter":

$$\mathbf{L} = \begin{pmatrix} -2.39E+1 & -2.39E+1 & -2.39E+1 \\ 6.09E-1 & 6.09E-1 & 6.09E-1 \\ -9.00E-3 & -9.00E-3 & -9.00E-3 \\ -2.24E-3 & -2.24E-3 & -2.24E-3 \\ 1.69E-3 & 1.69E-3 & 1.69E-3 \\ 6.08E+1 & 6.08E+1 & 6.08E+1 \\ -1.02E+1 & 1.02E+1 & 1.02E+1 \\ 1.49E-1 & 1.49E-1 & 1.49E-1 \\ 3.73E-2 & 3.73E-2 & 3.73E-2 \\ -2.68E-2 & -2.68E-2 & -2.68E-2 \end{pmatrix}. \quad (31)$$

The optimization was started with the foregoing observer gain matrix.
"vomosy":

$$\mathbf{L} = \begin{pmatrix} 0 & 4.56E+0 & 0 \\ 7.81E+0 & 0 & 4.07E-1 \\ -1.16E-1 & 0 & 0 \\ 0 & 0 & 0 \\ 0 & 0 & 0 \\ 0 & 0 & 3.13E+0 \\ 0 & 0 & 2.24E+0 \\ 0 & 0 & 0 \\ 0 & 0 & 0 \\ 0 & 0 & 0 \end{pmatrix}. \quad (32)$$

The following simulations present the output values, rigid body coordinate $\varphi(t)$, and the bending $v"(x_M, t)$, which is the consequence of the flexibility and the trajectory, respectively. The sign (-) denotes the ideal state feedback, while (- -) denotes the state feedback when the original states are substituted by the reconstructed states of the full-state observer.

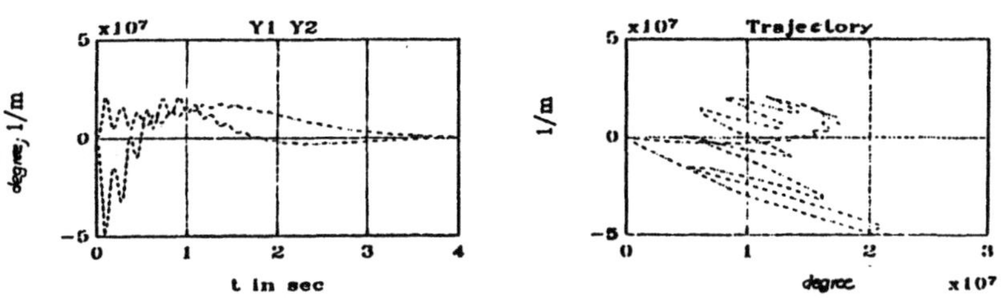

Figure 1: Ideal and observer feedback, "place"

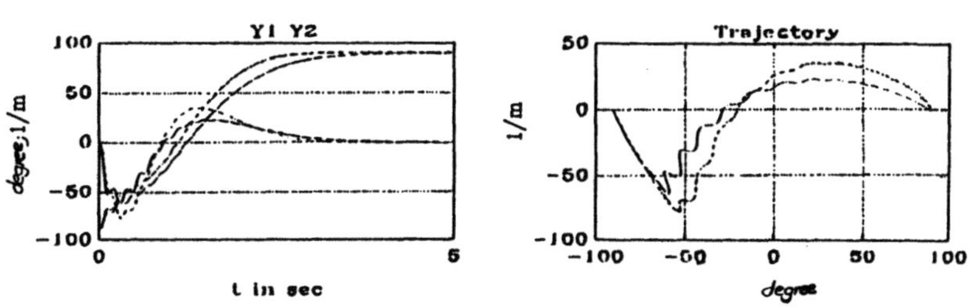

Figure 2: Ideal and observer feedback, "porter"

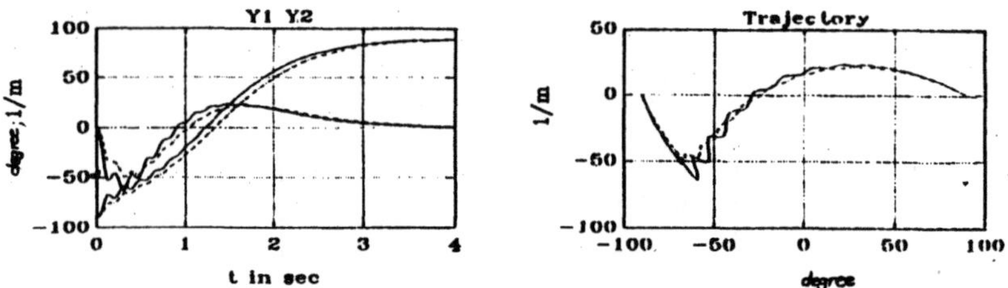

Figure 3: Ideal and observer feedback, "vomosy"

7. Minimal-state observer

If we choose the observer order ϑ in condition B_1 equal to $n - m$, then this leads to a minimal order state observer. According to Cumming [11] and Gopinath [12], the system (1), (2) can be transformed by the use of the regular coordinate transformation in such a way that the output matrix $\mathbf{C}$ has the following form.

$$\begin{aligned} \mathbf{x} &= \mathbf{G}\bar{\mathbf{x}}; \qquad \mathbf{G} = [\mathbf{g}_1 \quad \mathbf{g}_2] \\ \bar{\mathbf{C}} &= \mathbf{C}\mathbf{G} = [0 \quad \mathbf{I}_m] \end{aligned} \tag{33}$$

If

$$rg(\mathbf{C}) = m \,,$$

then $\mathbf{G}$ can be calculated by the Gaußianelimination procedure, or $\mathbf{H}_{(n-m)\times n}$ is chosen in such a way that

$$\mathrm{rg}\begin{pmatrix} \mathbf{H} \\ \mathbf{C} \end{pmatrix} = n, \text{ such that } \mathbf{G} = \begin{pmatrix} \mathbf{H} \\ \mathbf{C} \end{pmatrix}^{-1}$$

exists. The condition B_3 can be fulfilled if the transformation matrix $\mathbf{T}$ is defined as:

$$\mathbf{T} = [\mathbf{I}_{n-m} \quad -\mathbf{T}_2] \tag{34}$$

In the expression above, $\mathbf{T_2}$ is a free parameter which can be chosen arbitrarily. The coordinate transformation (33) partitions the original system (1) and (2) with

$$\bar{\mathbf{x}} = \begin{pmatrix} \bar{\mathbf{x}}_1 \\ \bar{\mathbf{x}}_2 \end{pmatrix}$$

into the new forms

$$\dot{\bar{\mathbf{x}}} = \begin{pmatrix} \bar{\mathbf{A}}_{11} & \bar{\mathbf{A}}_{12} \\ \bar{\mathbf{A}}_{21} & \bar{\mathbf{A}}_{22} \end{pmatrix} \bar{\mathbf{x}} + \begin{pmatrix} \bar{\mathbf{B}}_1 \\ \bar{\mathbf{B}}_2 \end{pmatrix} \mathbf{u}$$

$$\mathbf{y} = [0 \quad \mathbf{I}_m]\bar{\mathbf{x}} \,.$$

$\bar{\mathbf{A}}_{11} \in \mathrm{R}^{(n-m)\times(n-m)}, \quad \bar{\mathbf{A}}_{22} \in \mathrm{R}^{m\times m}$
If the transformation rule (34) is applied to the conditions B_2 to B_4, then the following

relations are valid between the system and the observer:

$$\mathbf{D}_{min} = \bar{\mathbf{A}}_{11} - \mathbf{T}_2\bar{\mathbf{A}}_{21} \tag{35}$$

$$\mathbf{L}_{min} = \mathbf{D}_{min}\mathbf{T}_2 + \bar{\mathbf{A}}_{12} - \mathbf{T}_2\bar{\mathbf{A}}_{22}$$

$$\mathbf{M}_{min} = \bar{\mathbf{B}}_1 - \mathbf{T}_2\bar{\mathbf{B}}_2 \,.$$

In summary the observer equations become

$$\dot{\mathbf{z}}(t) = \mathbf{D}_{min}\mathbf{z}(t) + \mathbf{L}_{min}\mathbf{y}(t) + \mathbf{M}_{min}\mathbf{u}(t)$$

$$\hat{\mathbf{x}}(t) = \mathbf{g}_1\mathbf{z}(t) + (\mathbf{g}_1\mathbf{T}_2 + \mathbf{g}_2)\mathbf{y}(t)$$

$$\mathbf{e}(t_0) = \mathbf{z}(t_0) - \mathbf{T}\mathbf{G}^{-1}\mathbf{x}(t_0) \,.$$

According to Cumming and Gopinath, the matrix pair $(\bar{\mathbf{A}}_{21}, \bar{\mathbf{A}}_{11})$ is completely observable if the matrix pair $(\mathbf{C}, \mathbf{A})$ is completely observable, too. Now the synthesis problem can be reformulated into a pole placement problem for the subsystem $\bar{\mathbf{A}}_{11}$, just as in the case of the identity observer. Hence the matrix $\mathbf{T}_2$ has to be chosen in such a way that $\mathbf{D}_{min}$ is a stable matrix within the meaning of condition B_5.

7.1 Limitation of the reconstruction error

In order to limit the reconstruction error, the gain matrix $\mathbf{T}_2$ must be designed in such a way that for the inequalities

$$\|\mathbf{e}(t)\| \le \|\mathbf{\Phi}_{Dmin}(t, t_0)\|\|\mathbf{e}(t_0)\| \tag{36}$$

$$\|\mathbf{e}(t_0)\| \le \|\mathbf{T}\|\|\bar{\mathbf{x}}(t_0)\|; \quad \mathbf{z}(t_0) \stackrel{!}{=} \mathbf{0} \tag{37}$$

$$\|\mathbf{L}_{min}\| = \|\mathbf{D}_{min}\mathbf{T}_2 + \bar{\mathbf{A}}_{12} - \mathbf{T}_2\bar{\mathbf{A}}_{22}\| \tag{38}$$

$$\le \|\mathbf{D}_{min}\|\|\mathbf{T}_2\| + \|\bar{\mathbf{A}}_{12}\| + \|\mathbf{T}_2\|\|\bar{\mathbf{A}}_{22}\| \tag{39}$$

the following conditions

$$\|\mathbf{\Phi}_{Dmin}(t, t_0)\| \to min; \qquad \text{for all } t \ge t_0 \,.$$

$$\|\mathbf{e}(t_0)\| \to min$$

$$\|\mathbf{L}_{min}\| \to min$$

have to be complied with, as well as condition B_5.

7.2 Main result for the closed loop system

Apart from the above-mentioned non-trivial problem, there exist a simple way of reducing the norm of the right-hand side of (36), (37) and (38). According to the foregoing section, the reconstruction error will also be limited. Therefore we make use of the remaining degrees of freedom when we generate the transformation matrix **G**. By assigning $\mathbf{H}_{(n-m)\times n}$, we only have to take into consideration that **G** becomes nonsingular. Together with equation (8), (35) and (34), we obtain for

$$\|\mathbf{T_2}\| \approx 0 \tag{40}$$

the approximation

$$\begin{aligned} \|\mathbf{\Phi}_{Dmin}(t,t_0)\| &\approx exp\|\bar{\mathbf{A}}_{11}t\| \qquad \text{for all } t \geq t_0 \\ \|\mathbf{L}_{min}\| &\approx \|\bar{\mathbf{A}}_{12}\| \\ \|\mathbf{e}(t_0)\| &\approx \|\bar{\mathbf{x}}(t_0)\| \,. \end{aligned}$$

Expression (40) can be achieved, if we choose the matrix $\mathbf{H}_{(n-m)\times n}$, which is a partial matrix of the coordinate transformation matrix **G**, in such a way that

$$\cdot\, \mathrm{rg}\begin{pmatrix} \mathbf{H} \\ \mathbf{C} \end{pmatrix} = n, \text{ such that } \mathbf{G} = \begin{pmatrix} \mathbf{H} \\ \mathbf{C} \end{pmatrix}^{-1}$$

exists and that as many eigenvalues as possible lie in the left half-plane for the subsystem $\bar{\mathbf{A}}_{11}$. The different solutions of the linear coordinate transformation by (33) can be used for certain classes of systems, so that the transformation yields a subsystem with all eigenvalues in the left half-plane. If $\bar{\mathbf{A}}_{12} \neq \mathbf{0}$, then a successful reconstruction in the sense of (6) and (14) can be achieved while $\mathbf{T_2}$ becomes zero. In many cases it is sufficient to move only the dominate eigenvalues of the subsystem somewhat further left into the left half-plane. Hence the norm of $\mathbf{T}_2$ can be kept small. If some of the produced eigenvalues lie in the saftey margin between the frequencies of the modelled and the unmodelled modes, they can be adjusted by a pole placement algorithm as shown in section . Then, $\mathbf{A} \equiv \bar{\mathbf{A}}_{11}^T$, $\mathbf{B} \equiv \bar{\mathbf{A}}_{21}^T$ and the solution $\mathbf{T_2} \equiv \mathbf{K}^T$.

8. Numerical example

The coordinate transformation for the minimal-order observer is realized by

$$\mathbf{G}_1 = \begin{pmatrix} 0 & 0 & 0 & 0 & 0 & 0 & 0 & 0 & 1 & 0 \\ 0 & 0 & 0 & 0 & 0 & 1 & 0 & 0 & 0 & 0 \\ 0 & 0 & 0 & 0 & 0 & 0 & 1 & 0 & 0 & 0 \\ -7.3211E-1 & 0 & 0 & 0 & 0 & -1.3580E-1 & -7.1015E-1 & 3.0469E-2 & 0 & 0 \\ 1 & 0 & 0 & 0 & 0 & 0 & 0 & 0 & 0 & 0 \\ 0 & 0 & 0 & 0 & 0 & 0 & 0 & 0 & 0 & 1 \\ 0 & 1 & 0 & 0 & 0 & 0 & 0 & 0 & 0 & 0 \\ 0 & 0 & 1 & 0 & 0 & 0 & 0 & 0 & 0 & 0 \\ 0 & 0 & 0 & 1 & 0 & 0 & 0 & 0 & 0 & 0 \\ 0 & 0 & 0 & 0 & 1 & 0 & 0 & 0 & 0 & 0 \end{pmatrix} \tag{41}$$

We obtain the following subsystem $\mathbf{A}_{11i}$ with some eigenvalues in the right half-plane. The new poles for $\mathbf{A}_{11i}$ are chosen according to [14] (see table 3). The corresponding gain matrix is

$$\mathbf{T}_2 = \begin{pmatrix} -3.66E-3 & -3.66E-3 & -3.66E-3 \\ 3.10E-1 & 3.10E-1 & 3.10E-1 \\ 5.63E-1 & 5.63E-1 & 5.63E-1 \\ 2.89E-1 & 2.89E-1 & 2.89E-1 \\ -3.51E-1 & -3.51E-1 & -3.51E-1 \\ 1.28E-2 & 1.28E-2 & 1.28E-2 \\ 4.90E-3 & 4.90E-3 & 4.90E-3 \end{pmatrix}. \tag{42}$$

The coordinate transformation

$$\mathbf{G}_2 = \begin{pmatrix} 0 & 0 & 0 & 0 & 0 & 0 & 0 & 0 & 1 & 0 \\ 0 & -5.2292E+0 & -7.3635E+0 & -5.3909E+0 & 0 & 0 & 0 & 2.2436E-1 & 0 & 0 \\ 0 & 1 & 0 & 0 & 0 & 0 & 0 & 0 & 0 & 0 \\ 0 & 0 & 1 & 0 & 0 & 0 & 0 & 0 & 0 & 0 \\ 0 & 0 & 0 & 1 & 0 & 0 & 0 & 0 & 0 & 0 \\ 0 & 0 & 0 & 0 & 0 & 0 & 0 & 0 & 0 & 1 \\ 0 & 0 & 0 & 0 & 0 & 1 & 0 & 0 & 0 & 0 \\ 0 & 0 & 0 & 0 & 0 & 0 & 1 & 0 & 0 & 0 \\ 1 & 0 & 0 & 0 & 0 & 0 & 0 & 0 & 0 & 0 \\ 0 & 0 & 0 & 0 & 1 & 0 & 0 & 0 & 0 & 0 \end{pmatrix} \tag{43}$$

yields a subsystem $\mathbf{A}_{11s}$ with (see table 3) all eigenvalues in the left half-plane. These eigenvalues are used for the observer. Thus we compute a gain matrix as

$$\mathbf{T}_2 = \begin{pmatrix} -1.77E-4 & -1.77E-4 & -1.91E-4 \\ -7.28E-6 & -4.53E-6 & 1.86E-6 \\ -8.79E-6 & 8.43E-6 & 2.03E-6 \\ 1.91E-5 & 2.02E-5 & -4.52E-5 \\ 5.01E-4 & 4.96E-4 & 5.02E-4 \\ -1.54E-4 & -1.59E-4 & -1.73E-4 \\ -3.31E-4 & -3.47E-4 & -3.54E-4 \end{pmatrix}, \tag{44}$$

so that $\mathbf{T}_2 \approx \mathbf{0}$.

subsystems			
$\mathbf{A}_{11i}$ unstable		$\mathbf{A}_{11s}$ stable	
Re	Im	Re	Im
-1.5310E+3	0.0	-1.4021E+3	0.0
-1.8722E+2	±1.0765E+2	-1.4166E+2	±1.3517E+2
-1.5936E+1	±2.2070E+1	-2.2773E+1	±7.6056E+1
+2.1779E+1	±8.9583E+1	-1.5421E+2	0.0
-.-	-.-	-8.6077E+0	0.0

Table 3: Eigenvalues of $\mathbf{A}_{11}$

The corresponding initial errors $\mathbf{e}(t_0)$ are shown in table 4. The following simulations

$\mathbf{e}(t_0)$	$\mathbf{e}(t_0)$
3.23E-1	-1.14E-2
-2.74E+1	4.38E-3
-4.99E+1	5.94E-4
-2.56E+1	-4.08E-3
3.10E+1	6.69E-3
-1.03E+0	5.84E-2
-4.31E-1	-7.13E-3

Table 4: Initial error

present the output values, rigid body coordinate $\varphi(t)$, and the bending $v"(x_M, t)$, which is the consequence of the flexibility and the trajectory, respectively. The sign (-) denotes the ideal state feedback, while (- -) denotes the state feedback when the original states are substituted by the reconstructed states of the minimal order-observer.

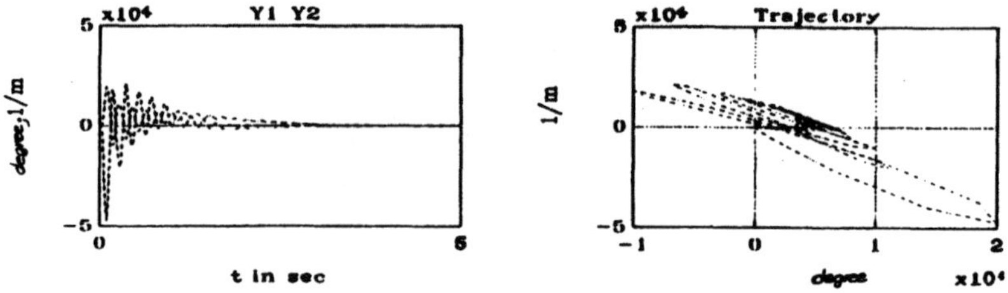

Figure 4: Ideal and minimal-observer feedback with $\mathbf{A}_{11i}$ unstable

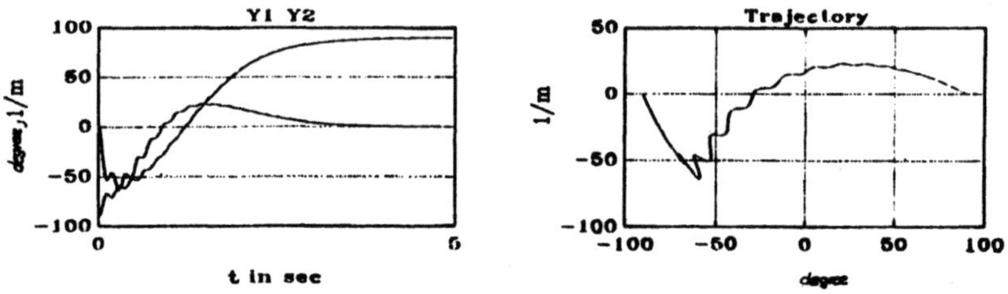

Figure 5: Ideal minimal- observer feedback with $\mathbf{A}_{11s}$ stable

9. Conclusion

By the use of well-known pole placement algorithms, especially by the modal approach of Roppenecker [4], a full-state observer and a minimal-state observer is designed for a flexible one-link robot arm. It is shown that the behaviour of the closed loop systems incorporating the state observers was improved if the norms of the gain matrices $\|\mathbf{L}\|$ and $\|\mathbf{T_2}\|$ were reduced. In the case of minimal-state reconstruction, we use the non-uniqueness of the coordinate transformation $\mathbf{G}$ to reduce the norm of the gain matrix of the minimal state observer $\|\mathbf{T_2}\|$ in a first step. Where necessary the above-mentioned pole assignment procedure is applayed in a second step.

References

[1] Truckenbrodt A. Bewegungsverhalten und Regelung hybrider Mehrkörpersysteme mit Anwendung auf Industrieroboter. Fortschritt-Berichte der VDI-Zeitschrift Reihe 8,No.33 (VDI-Verlag, Düsseldorf 1980)

[2] Ludyk G., Cordes J. Ahlers R., Bruce-Boye C., Fehren H., Schmidtke W. Analyse, Simulation und Regelung einer zeitveränderlichen, elastischen Struktur am Beispiel eines teleskopartigen Roboterarms. Teil 1. Modellbildung und Simulation eines elatischen Roboterarms. Berichte aus der Forschung der Universität Bremen (1988)

[3] Teil 2. Untersuchung der Beobachtbarkeit und Beobachterentwurf für einen elastischen Roboterarm. Berichte aus der Forschung der Universität Bremen (1988)

[4] Roppenecker G. Vollständige modale Synthese linearer Systeme und ihre Anwendung zum Entwurf strukturbeschränkter Zustandsrückführung. Fortschritt-Berichte der VDI-Zeitschrift Reihe 8, No.59 (VDI-Verlag, Düsseldorf 1983)

[5] Porter B., Crossley R. Modal Control, Theory and Applications. Taylor + Francis LTD London (1972)

[6] Kautsky,J., Nichols,N.K., van Dooren,P.: Robust pole assignment in linear state feedback. Int. J. Control Vol 41 (1985), S.1129-1155

[7] Luenberger G. D. An Introduction to Observers. IEEE Transaction on Automatic Control Vol AC-16, No.6, pp 596-602 (1971)

[8] Roppenecker G. Modal approach to parametric state observer design. Electronic Letters, Vol 22, No.3, pp 118-120 (1986)

[9] Roppenecker G. Entwurf robuster Regelungen mittels Vollständiger Modaler Synthese und ihre Anwendung auf den spurgeführten Omnibus. Automatisierungstechnik 35, S.349-358 (1987)

[10] Mensching S., Rohlfsen J. Entwurf eines robusten Reglers für den elastischen Roboterarm. Diplomarbeit an der Universität Bremen (1989)

[11] Gumming S.D. Design of observers of reduced dynamics. Electronics Letters, Vol 5, pp 213-214 (1969)

[12] Copinath B. On the control of linear multiple input-output systems. The Bell System Technical Journal, Vol 50, pp 1063-1081 (1971)

[13] Grübel G. Beobachter und Reglersynthese. Habilitationsschrift Ruhr-Universität Bochum (1977)

[14] Litz L., Preuss H.P. Ein Vorschlag zur Wahl der Pole beim Verfahren der Polvorgabe. Regelungstechnik, Vol 25, S.95-103 (1976)

[15] Wilkinson,J.H.:The algebraic Eigenvalue Problem. Oxford University Press (1965)

3

Model Building and System Simulation

Modelling and Simulation of Non-Isothermal Process for the Reduction, Precipitation and Characterization of Metal Phases in Zeolites

G. Schulz-Ekloff

1. Introduction

Chemical processes in advanced materials, like metal zeolites, proceed under influence of many parameters, like gas phase composition, temperature, metal type, ceramic stoichiometry, solid electrolyte acidity, degree of dehydration, etc. The process comprises many simultaneous or competitive reactions, e.g. metal reduction, diffusion, agglomeration, zeolite lattice fragmentation or reorganization, which can take place on different time scales and which often depend non-linearly on the temperature. The modelling of such complex systems can by no means be complete, i.e. cannot comprise all partial processes and corresponding parameters, but has to focus on the magnitudes mainly influencing a concerned phenomenon or property of the material. The elucidation of parameters of the first magnitude in non-linear systems can be facilitated by a technique studying transient states of relaxation or response following to a strong perturbation.

In the following the perturbation of Ni^{2+}, Co^{2+} and Pt^{2+} exchanged faujasites is achieved by applying linear temperature programs. The corresponding system responses are studied aiming to find out, which parameters dominate the obtained results, like degree of reduction (Co and Ni faujasites) or degree and type of metal dispersion (Pt faujasite).

2. Degrees of Reduction of Ni and Co Faujasites

The present chapter describes the temperature-programmed reduction (t.p.r.) of nickel and cobalt faujasites and the analysis of the non-isothermal kinetics, and reconsiders the reducibilities by making use of thermodynamic concepts. Sample preparation, characterization, apparatus and procedure have been described elsewhere [1]. Examples of t.p.r. spectra, obtained by differentiating the pressure p in the gas phase with respect to the temperature T, are given for NiX and CoX in figs. 2.1 and 2.2. The t.p.r. spectra were analysed by setting up second-order kinetics for the reduction process according to

(2.1) $Ni^{2+} + H_2 = Ni^{o} + 2H^{+}$.

The assumption of a reduction process of second order, and the application of a linear heating rate $\beta = dT/dt$ (K min^{-1}) leads to the following expression for the kinetics:

(2.2) $\quad - zdp = kzpq \; dT/\beta$

including the hydrogen pressure, p, the number of moles of hydrogen, zp, the number of moles of reducible cations, q, and the rate constant, k. $z = V_{app}/RT$, including the volume of the closed-circuit apparatus V_{app} and the

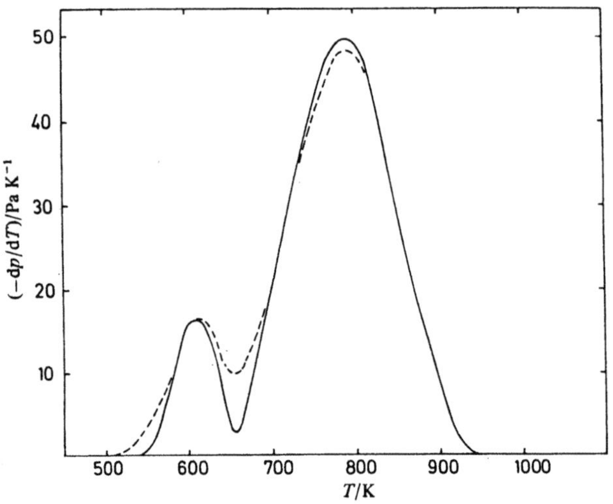

Fig. 2.1: Temperature-programmed reduction of NiX by hydrogen (full line) and its simulation (broken line).

gas constant R. $k = k_0 exp(-E/RT)$, including the frequency factor k_0 and the activation energy E, as usual. The good resolution of the two peaks in the t.p.r. spectra (fig. 2.2 and 2.3) enables the two reduction steps to be considered as separate processes, taking

(2.3) $\quad -dp = k_1 zp(p_1 - p_{1,2})dT/\beta$

and

(2.4) $\quad -dp = k_2 zp(p_{1,2} - p_2)dT/\beta$

including the initial pressure, p_1, the final pressure for the first reduction step, $p_{1,2}$, and the final pressure for the second reduction step, p_2. The number of cations q reduced in each step is balanced by $q_1 = z(p_1 - p_{1,2})$ and $q_2 = z(p_{1,2} - p_2)$. Since the pressure values in the domain of

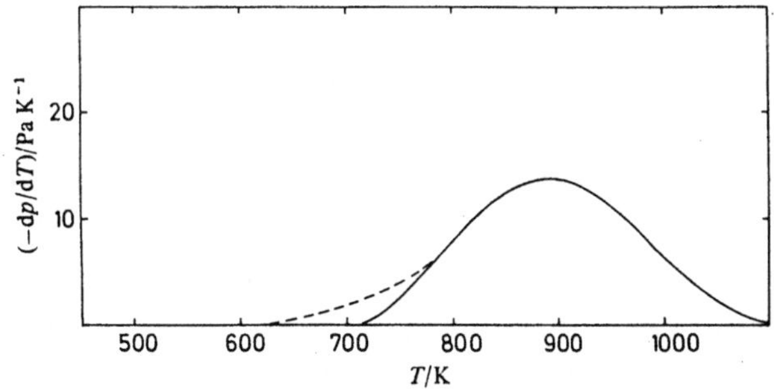

Fig. 2.2: Temperature-programmed reduction of CoX-PtX mixture by hydrogen (full line) and its simulation (broken line).

overlapping exhibit larger errors the values of $p_{1,2}$, used for the model, were estimated by the following iterative calculation. Arrhenius parameters k_{o2} and E_2 were determined from the rearranged and linearized equation (2.4)

(2.5) $\ln[1/p(p^*_{1,2}-p_2)(-dp/dT)] = \ln(k_{o2}z/\beta) - E_2/RT$

using the erroneous experimental pressure $p^*_{1,2}$, i.e. at the minimum between the peaks, for the second reduction step. Values for $p_{1,2}$ were then calculated by integrating eqn (2.4) within the limits $(p_{1,2},p)$ and (T_o,T)

(2.6) $(p-p_2)/p = \alpha_2\exp[-a_2I_2(T)]$

including

(2.7) $\alpha_2 = (p_{1,2}-p_2)/p_{1,2}$

(2.8) $a_2 = k_{o2}zp_2/\beta$

(2.9) $I_2(T) = \int_{T_o}^{T} \exp[(-E_2/RT)dT]$

$I_2(T)$ was solved numerically and the experimental data were taken for the calculation of $(p-p_2)/p$. The value $p_{1,2}$ obtained from eqn (2.6) was used for the evaluation of E_1, E_2, k_{01} and k_{02} (fig. 2.3). These data were used for simulating the t.p.r. spectra (fig. 2.1–2.3). For this simulation, the derivations dp/dT were used, e.g. for the second peak from eqn (2.6).

(2.10) $(dp/dT)_2 = -\alpha_2 a_2 p_2 \exp\{-a_2 I_2(T)\}\exp(-E_2/RT)/(1-\alpha_2\exp\{-a_2 I_2(T)\})^2$

The estimated $p_{1,2}$ values were used for the calculations of the fractions of cations reduced in each step, i.e. $x_1 = (p_1-p_{1,2})/(p_1-p_2)$ and $x_2 = (p_{1,2}.p_2)/(p_1-p_2)$. A peak analysis can be carried out by determining the fraction of reducible cations at the peak maximum, x_m. The procedure is described for the second peak as an example. The derivation of eqn (2.10), $(d^2p/dT^2)_2$, and the condition at the maximum, $(d^2p/dT^2)_{m2} = 0$, give

(2.11) $(a_2 R T_{m2}^2/E_2)\exp(-E_2/RT_{m2})=(1-\alpha_2\exp\{-a_2 I_2(T_{m2})\}/(1-\alpha_2\exp\{-a_2 I_2(T_{m2})\})$

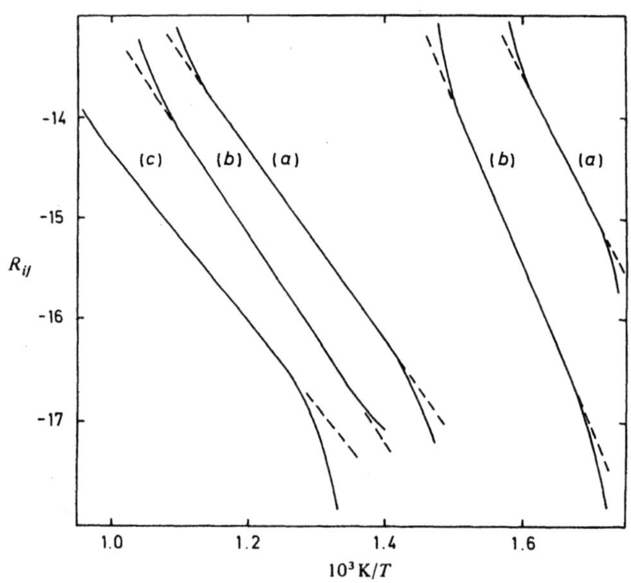

Fig. 2.3: Linearised functions for the measured (full lines) and simulated (broken lines) kinetics of the temperature-programmed reduction: (a) NiX, (b) NiY and (c) CoX/PtX. $R_{ij} = \ln\{[1/p(p_i-p_j)](-dp/dT)\}$; $p_i \equiv p_1, p_{1.2}$; $p_j \equiv p_{1.2}, p_2$.

From the experimental condition it follows that $\alpha < 0.2$. The condition $a_2I_2(T) > 0$ gives $\exp[-a_2I_2(T)] < 1$, and the right-hand side of eqn (2.11) tends to 1. It follows that

(2.12) $\exp[-E_2/RT_{m2}] = E_2/a_2RT^2_{m2}$

The integral $I_2(T)$ is expressed by an approximation with sufficient accuracy as

(2.13) $I_2(T) = R/E_2[T^2\exp(-E_2/RT)-T^2_0\exp(-E_2/RT_0)]$

For $T = T_m$ eqn (2.12) and (2.13) combine to give

(2.14) $I_2(T_{m2}) = (1/a_2)-(RT^2_0/E)\exp[-E_2/RT_0]$

For $T_0 = 293$ K the second term on the right-hand side of eqn (2.14) is negligible compared with $1/a_2$ giving

(2.15) $I_2(T_{m2}) = 1/a_2$.

Substitution of this value for $I_2(T_{m2})$ into eqn (2.6) gives

(2.16) $(p_{m2}-p_2)/p_{m2} = \alpha_2\exp(-1)$.

The combination of eqn (2.16) and (2.7) with $x_{m2} = (p_{m2}-p_2)/(p_{1,2}-p_2)$

(2.17) $x_{m2} = (1-\alpha_2)\exp[-1]/1-\alpha_2\exp[-1]$

For a surplus of hydrogen over nickel cation, i.e. for low values of $_2$, the value of x_{m2} approaches $\exp(-1) = 0.37$.

In general, the reducibility of cation-exchanged zeolites is defined by the degree of reduction obtained in any standardised experiment. The phenomenon of incomplete reduction was attributed to the existence of redox equilibria in zeolites, which can be considered as solid solutions [3]. This means that the electron transfer always proceeds as a reversible step, whereas the overall reduction is finally irreversible, owing to subsequent irreversible processes.

The most important subsequent reaction is the metal agglomeration. However, the increase in the proton activity a_{H^+} [eqn(2.1)] can also cause irreversible subsequent processes, such as dehydration or dealumination. The redox process can be described by the chemical potential μ:

$$(2.18)\quad \mu = \mu_{o,Ni} + RT\ \ln[(1-n)a_{Ni^{2+}}/(a_{H^+}+2na_{Ni^{2+}})^2]$$

where a_{H^+} is the proton activity prior to reduction, $a_{Ni^{2+}}$ is the nickel ion activity and n is the fraction of the reduced metal ions. $\mu_{o,Ni}$ contains a given constant partial pressure of hydrogen. The activity of the metallic nickel is assumed to be $a_{Ni,o} = 1$. From eqn (2.18) it follows that a redox system μ_2, differing from a reference system μ_1 only by higher initial proton activity a_{H^+}, will arrive at the degree of reduction of the reference system at higher temperature according to

$$(2.19)\quad T_2/T_1 = \ln(a_{H^+})_1/\ln(a_{H^+})_2.$$

For example, the difference in temperatures T_2-T_1 will be in the range 50–100 K from a ratio $(a_{H^+})_2/(a_{H^+})_1=2$ at an order of magnitude for a_{H^+} of $10^{-3}-10^{-4}$ mol g^{-1}. This means that the relative reducibilities can be defined either by the degree of reduction at a constant temperature or by the tempe-rature necessary for a constant degree of reduction. The fraction of unreduced but reducible cations $x_m = (p_m-p_f)/(p_i-p_f)$, where p_i, p_f and p_m are the pressures of hydrogen initially, finally and at the peak maximum, was found to be constant at the maximum T_m of the t.p.r. spectrum, i.e. $x_m \approx 0.4$, in all cases. This means that the T_m values represent relative reducibilities and that the experimental x_m values obtained are in accordance with the applied kinetics [eqn (2.17)].

3. Reduction-Agglomeration Model of Pt-Faujasite

In the following a reduction-agglomeration model is proposed, which simulates the experimentally obtained metal-particle size distribution in reduced platinum-exchanged NaX zeolite, considering the influence of the auto-reduction media, the platinum loading and the degree of dehydration [4].

The temperature-programmed autoreduction of platinum tetra-ammine faujasite X is monitored by the thermal desorption of nitrogen. Only one nitrogen peak is found if the reduction is carried out in a gaseous medium such as argon or oxygen. The nitrogen-evolution curve can exhibit two maxima, if the reduction proceeds in a vacuum (fig. 3.1[a]).

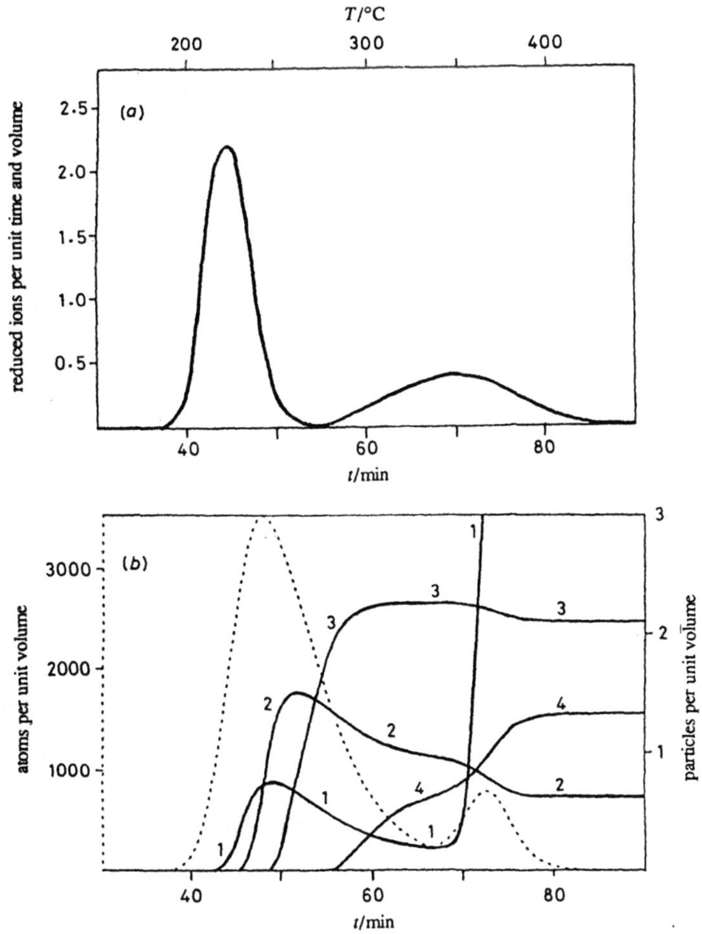

Fig. 3.1[a]: Nitrogen evolution in the temperature-programmed decomposition of platinum tetra-ammine-ion-exchanged NaX. The ordinate values are converted into reduction rates used for the simulation. The maxima correlate with the reduction of mononuclear (ca. 500 K) and polynuclear (ca. 630 K) complexes; [b] simulated rate of the change in the supply of atoms (including clusters) by the reduction and the agglomeration (···); changes in the relative fraction of particles of selected size (integer nm) (———).

The electron micrographs showed uniform distributions of the metal phase following the temperature-programmed decomposition of the platinum-tetra-ammine-exchanged zeolite NaX. This is also valid for the platinum particles

exceeding the supercage size as has been demonstrated in a previous paper [5]. The particle size distributions were evaluated from sets of micrographs taken from different specimens of the same sample charge. The bars in the histograms of fig. 3.2 represent integer numbers.

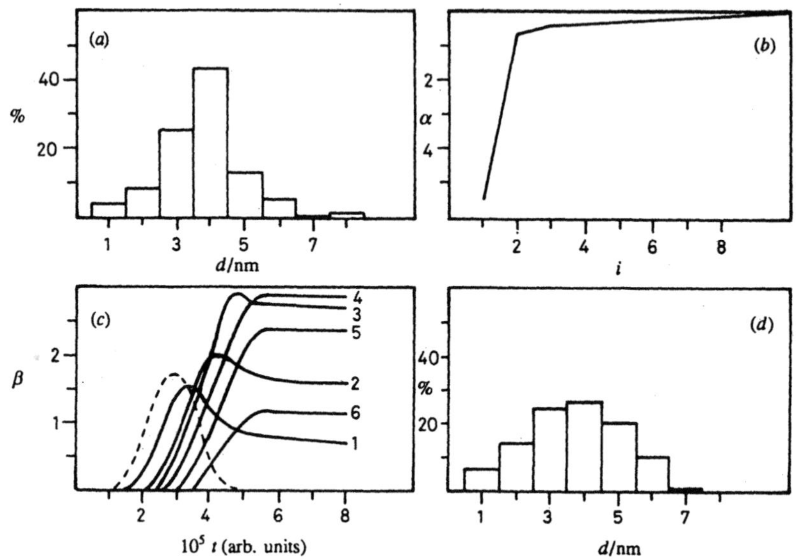

Fig. 3.2: (a) Histogram of the experimental metal dispersion; (b) fitted sticking probability (α = −log H) as a function of the number of atoms i in an agglomerate; (c) simulated change of the metal-atom concen-tration (β = atoms/10^3 per unit volume) with time t (arb. units) (---); simulated change in the relative fraction of particles of selected size (nm) with time (arb. units) (———); (d) histogram of the simulated metal dispersion.

Assuming that the autoreduction processes are thermally activated, the reduction rate at constant heating rate can be described by a differential equation of the form

$$(3.1) \quad R(t) = -\, dZ^{2+}/dt = Z^{2+}(t)P\exp[-B/a(t_0+t)]$$

where the temperature T is replaced by the time t. The constant a is the heating rate and $Z^{2+}(t)$ is the remaining concentration of platinum complex ions still available for the reduction process at time t. P, B and t_0 are fit parameters considered to be constant, i.e. not dependent on the temperature

and valid for all autoreduction processes observed. The values of the parameters were determined by fitting R(t) to the experimental nitrogen desorption curve. Two eqn (3.1) were superimposed in the case of nitrogen-desorption curves exhibiting two maxima. Following the determination of $Z^{2+}(t)$ by numerical integration, the rate R(t) of atom production as a function of time can be calculated. Metal atoms formed in the autoreduction step of the platinum tetra-ammine decomposition process can diffuse through the zeolite framework and collide with other metal atoms, clusters or crystals. The diffusion of metal atoms within the faujasite structure is three-dimensional and non-activated owing to weak interactions with the zeolite lattice [6]. The collision of a platinum atom with another platinum species (atom, cluster or crystal) can, in all cases, result in agglomeration, since the heterogeneous nucleation of platinum does not require the formation of a nucleus of critical size [7]. The aggregation probability of a collision will depend only on the possibility of accommodating the collision impulse, e.g. by excitation of lattice vibrations. The probability of capturing a colliding atom will increase with increasing size of the collision partner and will be close to unity for larger metal clusters. This agglomeration model considers only the capture of diffusing and colliding atoms. Sintering via diffusion and coalescence of clusters or small metal particles is neglected in view of the temperature and time scales considered. Sintering via thermal evaporation of metal atoms from agglomerates is neglected with respect to the high heat of vaporization (ca. 450 kJ mol^{-1}) [8] and the high supersaturation. The ratio of the platinum-atom concentration Pt^0 resulting from the reduction rate and the platinum concentration in the saturation equilibrium Pt_s^0 [9] is of the order of magnitude of $Pt^0/Pt_s^0 \approx 10^5$. The number of particles containing i atoms which can add an atom per unit volume and time is given by

$$(3.2) \quad A_i(t) = GQ_{1i}Z_1(t)Z_i(t)F$$

where G is a collision frequency factor, Q_{1i} the effective collision cross-section for a single atom with a particle of i atoms, defined by $Q_{1i}= (d_1+d_i)/2$ with d_i the diameter of particles containing i atoms. H_i is the sticking or capture probability in a collision of one atom with a particle of i atoms. Z_i is the number of particles per unit volume containing i atoms. F is a factor regarding the growth of particles beyond supercage dimensions. F = 1 for particles below and F < 1 for particles beyond supercage dimension, depending in the degree of dehydration.

The differential equations Z_i are

(3.3) $dZ_1/dt = R(t) - A_1(t)$

(3.4) $dZ_i/dt = A_{i-1}(t) - A_i(t) \quad (i > 1)$

Temperature-dependence of the collision frequency factor was neglected. H_i = 1 was set for $i > 10$ and a function

(3.5) $H_i = \exp[-x/(i+y)]$

was used for values of i from 1 to 9. For the autoreduction process involving polynuclear complexes in the supercages, i represents the initial number of platinum atoms per complex. Eqn (3.3) and (3.4) were integrated numerically. The values of the unknown parameters G, x and y for the dynamic growth were obtained by fitting the calculated final particle-size distribution to the experimental one. It turned out that an identical set of these parameters was valid for most of the simulations matching the experimental results. Modifications could be related to corresponding modifications in the mechanism. The predictive character of the model was proved by simulating and measuring intermediate particle size distributions [10]. For the simulation a unit volume of 8×10^3 nm^3 (10^3 unit cells) and particle size diameters d ≤ 8 nm ($i \leq 10^4$) were considered. The program was written in FORTRAN 77 and implemented on a Siemens 7880 computer.

4. Thermal Analysis of Pt Dispersions by O_2 as Probe Molecule

The temperature-programmed desorption (TPD) was carried out in a stainless steel HV-chamber equipped with a quadrupole mass spectrometer (Q200, Leybold-Heraeus), a turbo molecular pump (Turbovac 220, Leybold-Heraeus), an ionisation gauge and a controlled leak valve. The base pressure was about 10^{-6} Pa. The instrument was calibrated with oxygen to convert the partial ion current into the partial pressure. The electrical furnace inside the HV-chamber could be heated up to 893 K. The heating rate was in general 5.75 K/min. The sample (5 - 15 mg) was introduced into the furnace in the HV-chamber via a first vacuum chamber equipped with a gas inlet valve. To ensure that no loss of the sample occurred during the transfer, the weight was controlled before and after the experiment.

The mass spectrometer, thermoelement and ionisation gauge were connected with a PDP11-computer (DEC). Every minute a complete mass spectrum in the range of 1 to 60 amu was recorded, so it could be checked wether further gases desorb from the sample. Besides an unavoidable water desorption from the zeolite and the normal background in this vacuum range no further substances were detected. The desorption curve was evaluated and smoothened by a cubic spline function. In a first temperature desorption the samples stored in air were heated up to 893 K, thereby removing the main part of the water at the same time. The vacuum reduced samples were used directly after reduction in the HV-chamber. The subsequent TPD cycles for the sample reduced in Ar and O_2 and the first TPD cycle for the vacuum reduced samples were carried out as follows. After cooling down to room temperature the Pt-catalysts were transported into the first vacuum chamber and there exposed to dried oxygen at 10^5 Pa for 1 h. Thereafter they were transferred back into the HV-chamber and kept there for 2 h until the pressure was lower than 4×10^{-6} Pa before starting the TPD cycle. With this adsorption condition equal coverage for all samples was assured.

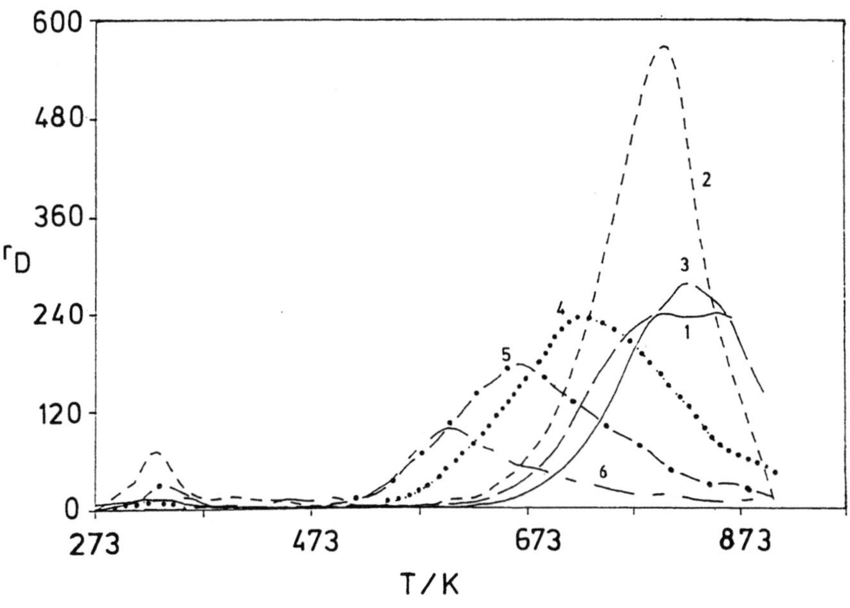

Fig. 4.1: Oxygen desorption from platinum single crystals with a surface average diameter ranging from 1-5 nm. The desorption peak shifts to higher temperature with decreasing particle size. r_D: desorption rate in 10^{-6} mol/mol Pt; 1: Pt12X/O_2; 2: Pt17Y/O_2; 3: Pt42X/V; 4: Pt12X/V; 5: Pt17Y/Ar; 6: Pt12X/Ar.

The temperature-programmed desorption spectra for oxygen obtained for various samples [11] are plotted in fig. 4.1. No oxygen desorption could be observed from a blank NaX sample. The position of the desorption maximum and the shape of the curves vary in dependence on the catalyst in the temperature range between 500 and 900 K. Residual water still adsorbed on the Pt-zeolite has no effect on the shape of the desorption spectrum or on the metal particle size distribution following the desorption cycle. This could be demonstrated by adsorption experiments with oxygen saturated with water at ambient temperature in agreement with observations that molecular bound water desorbs from Pt (111) and (110) surface already between 100 and 200 K [12,13]. Oxygen evolving from the catalytic decomposition of water can be excluded since no hydrogen could be detected. The nearly linear shift of the maximum of the oxygen desorption to lower temperatures with increasing surface average diameter of the platinum crystals is depicted in Fig. 4.2.

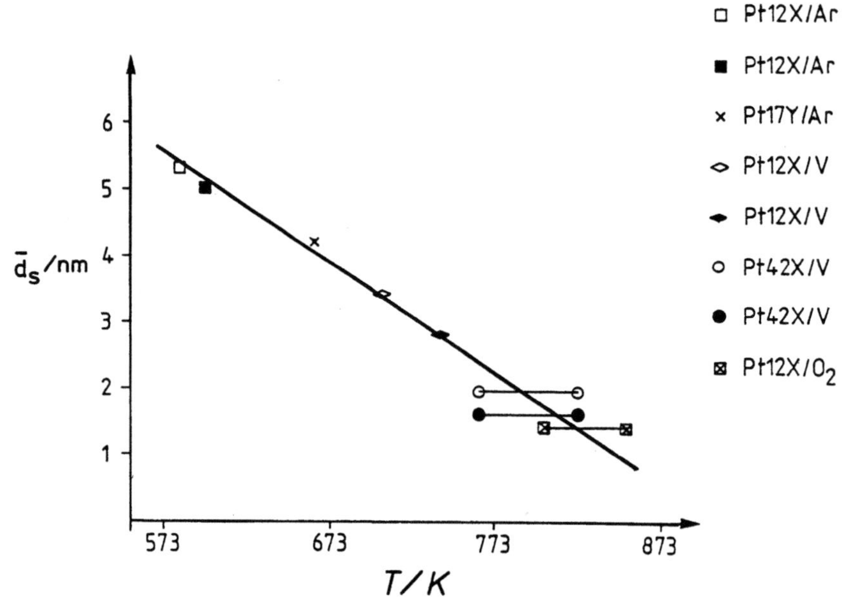

Fig. 4.2: Plot of the surface average diameters versus the desorption peak maxima. Pt12X/O_2 and Pt42X/V show two desorption peaks. The range from the first to the second maximum is marked, respectively.

The modelling of the oxygen desorption considers in a first approximation the pressure increase by desorption

(4.1) $dp/dt = (RT_R/V_R)\cdot(dn_{des}/dt)$

using the pressure p, the time t, the recipient temperature T_R, the recipient volume V_R, the gas constant R and the moles of desorbed oxygen n_{des}.
The pressure drop at a given time t by pumping is

(4.2) $dp/dt = -(S/V_R)\cdot p(t)$

using the pumping rate S. Combining (4.1) and (4.2) gives the change of the recipient pressure p_R

(4.3) $dp_R/dt = (dn_{des}/dt)(RT_R/V_R)-(S/V_R)p(t)$

or the desorption rate at a given time t

(4.4) $dn_{des}/dt = (V_R/RT_R)(dp_R/dt) + (S/RT_R)p(t)$

Since S is much larger than dp_R/dt, a reduction of (4.4) to

(4.5) $dn_{des}/dt = (S/RT_R)p(t)$

can be carried out. With a linear heating rate $dT/dt = \beta$ follows

(4.6) $dn_{des}/dT = (S/RT_R\beta)\ p(T)$

Integration of (4.6) yields

(4.7) $n_{des}(T) = (V_R/RT_R)p(T) + (S/RT_R\beta) \int_{T_0}^{T} p(T)dt$

For the activated desorption kinetics holds

(4.8) $dn_{des}/dT = (k_0/2\beta)(n_{ad}(T))^2\exp(-Ea/RT)$

using the moles of adsorbed oxygen atoms n_{ad}, the activation energy Ea, the frequency factor k_0 and the oxygen molecule dissociation $n_{des} = 2n_{ad}$.
Linearization of (4.8) using the logarithm gives

(4.9) $\ln(dn_{des}/dT)-2\ln\ n_{ad}(T) + \ln(2\beta) = -Ea/RT + \ln k_0$

n_{ad} can be calculated from (4.7) using

$$(4.10)\quad n_{ad}(T) = n_{max} - 2n_{des}(T) = 2n_{des,max} - 2n_{des}(T)$$

resulting in

$$(4.11)\quad n_{ad}(T)=(2V_R/RT_R)[p(T_{max})-p(T)]+(2S/RT_R\beta)\cdot\left(\int_{T_0}^{T_{max}} p(T)dT - \int_{T_0}^{T} p(T)dT\right)$$

Combination of (4.9) and (4.11) allows the simulation of the experimental thermal spectra by appropriate choice of the chemical parameters Ea and k_0. It turned out that the experimental spectra could be fitted best by assuming values for the activation energy Ea, which increase with decreasing values $n_{ad}(T)$.

5. Conclusions

The thermal analysis revealed that the non-isothermal reduction of nickel and cobalt faujasites can be described quantitatively by a model of kinetics based on an irreversible overall reaction of second order. The application of thermodynamic concepts enables the reducibility in a non-isothermal reduction to be defined by the position of the peak maximum of the t.p.r. spectrum. Relative reducibilities are related to proton activities or overall acidities. The agreement between the dynamics of the simple collision-agglomeration model proposed for the platinum precipitation in the zeolite NaX and the experimental size distribution is satisfactory.

Zeolite supported platinum phases which exhibit the structural speciality that the metal crystallites are uniformly surrounded by the zeolite support show a particle size effect in the chemisorption of oxygen. The shift of the desorption curves to higher temperatures with decreasing metal particle size is interpreted in a first approach with a corresponding increase of the surface free energy.

Deeper insights into the complex mechanisms of the chemical processes could be obtained from the simulation models which enabled the appropriate parameter variations to find out the magnitudes predominantly influencing the conversion of matter and, thus, to establish the final correlations for the mathematical models describing the process phenomenons.

Acknowledgement

Figures 2.1 - 2.3 were taken from reference [1] and figures 3.1 and 3.2 from reference [4] by courtesy of The Chemical Society.

References

[1] G. Schulz-Ekloff, L. Czarnetzki and A. Zukal, J. Chem. Soc., Faraday Trans. 1, 83 (1987) 3015.

[2] N.W. Hurst, S.J. Gentry, A. Jones and B.D. McNicol, Catal. Rev. Sci. Eng. 24 (1982) 233.

[3] L. Riekert, Ber. Bunsenges. Phys. Chem. 73 (1969) 331.

[4] D. Exner, N. Jaeger, A. Kleine and G. Schulz-Ekloff, J. Chem. Soc., Faraday Trans. 1, 84 (1988) 4097.

[5] A. Kleine, P.l. Ryder, N. Jaeger and G. Schulz-Ekloff, J. Chem. Soc., Faraday Trans. 1, 82 (1986) 205.

[6] J. Sauer, H. Haberlandt and W. Schirmer, in "Structure and Reactivity of Modified Zeolites", ed. P.a. Jacobs, N.I. Jaeger, P. Jiru, V.B. Kazansky and G. Schulz-Ekloff (Elsevier, Amsterdam 1984); Stud. Surf. Sci. Catal., vol. 18, p. 313.

[7] P. Wynblatt, in "Growth and Properties of Metal Clusters", ed. J. Bourdon (Elsevier, Amsterdam 1980); Stud. Surf. Sci. Catal., vol. 4, p. 15.

[8] Handbook of Chemistry and Physics, ed. R.C. Weast (Chemical Rubber Co., Cleveland, 1974).

[9] A.N. Nesmeyanov, Vapor Pressure of the Chemical Elements (Elsevier, Amsterdam 1963).

[10] N.I. Jaeger, A. Jourdan, G. Schulz-Ekloff, A. Kleine and P.L. Ryder, in "Innovation in Zeolite Materials Science", ed. P. Grobet, W.J. Mortier, E.F. Vansant and G. Schulz-Ekloff (Elsevier, Amsterdam 1988); Stud. Surf. Sci. Catal., vol. 37, p. 191.

[11] N.I. Jaeger, A.L. Jourdan, G. Schulz-Ekloff and A. Svensson, in "Structure and Reactivity of Surfaces" ed. C. Morterra, A. Zecchina and G. Costa (Elsevier, Amsterdam 1989); Stud. Surf. Sci. Catal., vol. 48, p. 503.

[12] F.T. Wagner and T.E. Moylan, Surface Sci. 191 (1987) 121.

[13] J. Fusy and R. Ducros, Surface Sci. 176 (1986) 157.

Dynamics of Heterogeneously Catalyzed Oxidation Reactions on Palladium Supported Catalysts

N.I. Jaeger

Abstract The complicated dynamic behavior of two heterogeneously catalyzed reactions, the complete oxidation of ethanol and the oxidation of carbon monoxide is analyzed and discussed. Two approaches for the development of mathematical models based upon the experimental results are presented.

1. Introduction

Heterogeneously catalyzed chemical reactions are at the heart of most industrial chemical processes and the field of heterogeneous catalysis has a long history of successful and ever intensifying basic research [1]. In recent years the study of complicated dynamic behavior, e.g. limit cycle or chaotic oscillatory behavior and multistability, namely in heterogeneous catalytic oxidation reactions has attracted considerable attention [2,3] following the observation and interpretation of oscillations in the oxidation of carbon monoxide on platinum supported catalysts by Beusch et al. [4]. The analysis of dynamic phenomena like the formation of temporal and spatial order in open chemical reaction systems far from equilibrium led to new insights into basic mechanisms of heterogeneous catalysis. A comprehensive review of recent experimental and theoretical developments can be found in [5]. Oscillations in heterogeneous catalytic reactions are normally observed on a macroscopic scale either as product oscillations in the outlet of the reactor or as oscillations of the average temperature of the catalyst bed. As a prerequisite for these observations the coupling of a large number of active centers, e.g. noble metal particles a few nanometers in diameter on a ceramic support, is required. The experimental phenomena therefore have to be explained on the basis of chemical kinetics by local oscillators or multistable centers which then have to be coupled by diffusive processes across a catalytically active surface.

The complicated dynamic behavior of two reactions, the complete combustion of ethanol and the oxidation of carbon monoxide to carbon dioxide will be discussed and two different approaches for mathematical modelling will be presented.

2. Experimental procedure

As catalysts served dispersions of palladium on amorphous aluminum oxide or within a zeolite matrix. The reactions were carried out in continuous flow reactor systems. As an example the reactor used in the oxidation of ethanol is shown in Figure 1.

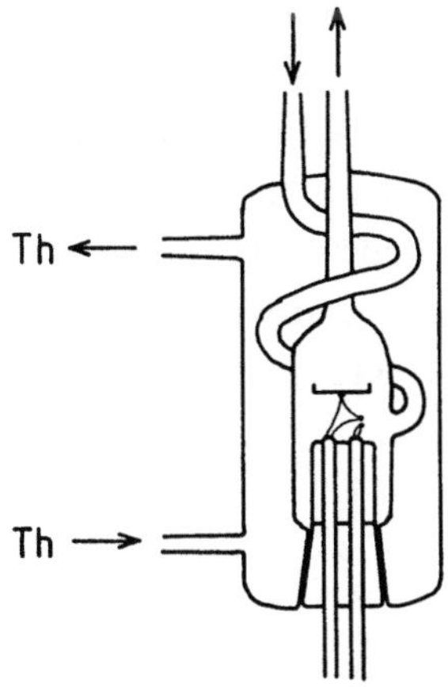

Figure 1: Continuous flow reactor for calorimetric measurements [6].

It consisted of a small reaction cell which was surrounded by a heatable glass tube (Th). The catalyst itself was placed on a silver plate which was connected to a thermocouple. The difference in temperature (ΔT) between this silver plate and the reaction gas at the reactor inlet served as a direct measure of the reaction rate of the complete combustion of ethanol and was continuously registrated.

The analysis of the reaction products was carried out by means of mass spectroscopy or by continuously monitoring the CO or CO_2 concentrations in the outlet of the reactor by nondispersive infrared absorption spectroscopy. It is essential in the experiments that flow rates, feed concentrations and the temperature of the reactor are precisely controlled. In addition the reactor has to be characterized to assure that the dynamic features observed are within the kinetic regime of the reactions.
Experimental details are given in [7,8] for the oxidation of ethanol and in [9,10] for the oxidation of CO.

3. The oxidation of ethanol

Dynamic systems can be characterized by the experimental determination of bifurcation diagrams which describe the response of the system to a slowly changing operating variable , e.g. the reaction temperature or a concentration. The identification of experimentally observed bifurcations, i.e. transitions in the dynamic behavior, can be used for the formal construction of a mathematical model. The model has to account for the correct sequence of observed transitions and divides the parameter space into regions with qualitatively different types of behavior. This has been shown by Sheintuch and Luss [11,12] for a system with two dynamic variables which have widely separated time scales. In the oxidation of ethanol only steady states and periodic oscillations of the reaction rate could be observed. The shape of the corresponding ΔT-curves is signified by the feature of relaxation oscillations [13]. Figure 2 gives an example: a steeply rising flank corresponding to a sharp increase in the reaction rate is followed by a relatively slow decrease of ΔT.

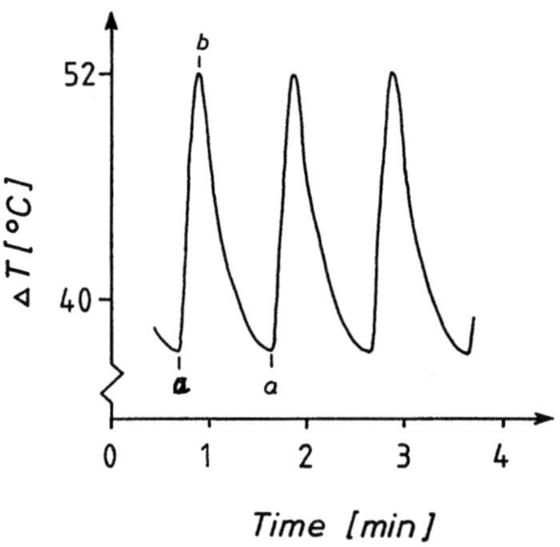

Figure 2: Typical shape of temperature oscillations

The shape of the oscillations is taken as evidence for the existence of sufficiently separated time scales. This allows the description of the oscillations by a deterministic two dimensional dynamical system

$$\varepsilon\dot{x} = f(x,y,\vec{k})$$

$$\dot{y} = f(x,y,\vec{k})$$

with one slowly and one rapidly changing variable y and x respectively ($\varepsilon \to 0$). $\vec{k}$ is the vector of external parameters.

Figure 3 presents a bifucation diagram which describes the influence of the oxygen content upon the reaction rate. All other experimental parameters were kept constant. Stable steady states are represented by full lines, stable oscillatory behavior is symbolized by vertical lines the length of which represents the amplitude of the oscillations.

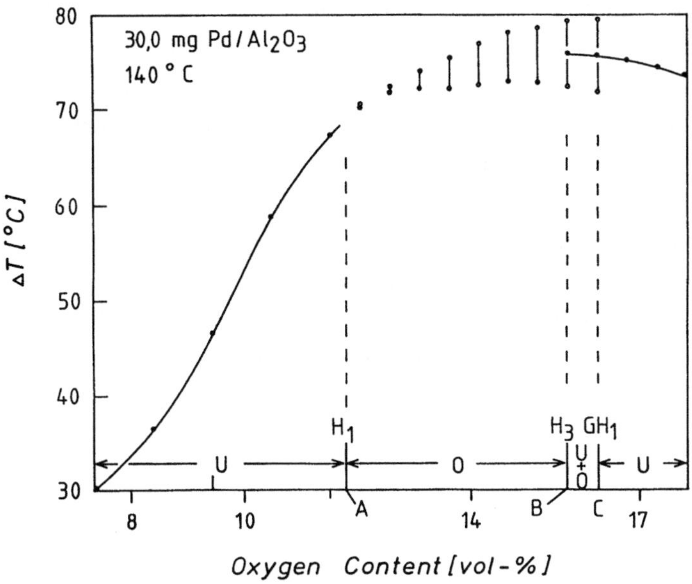

Figure 3: Bifurcation diagram for the palladium catalyzed oxidation of ethanol

At point A a transition of a stable steady state (U) to a stable oscillatory state (O) via a Hopf-bifurcation of type H_1 can be identified (see [11] for the notation). At point C a hard transition leads to a single stable steady state. This state is found to overlap the oscillatory state if the diagram is entered from the right hand side. The region of coexisting states in phase space is denoted by (U+O). The actually observed state depends on the direction of change of the control parameter. The occurence of this region (U+O) for which a stationary state lies within the amplitude of an oscillatory state is found next to a Hopf-bifurcation of type H_3 starting with a stable oscillation or next to a global Hopf-bifurcation of type GH_1 if the region is entered from the stable steady state. All the identified bifurcations and the overlapping state (U+O) can be obtained in the correct sequence by two differential equations [14].

$$\varepsilon\dot{x} = -x^5 + k_5x^3 + k_4x - y$$

$$\dot{y} = k_2x^2 + k_1x - y - k_3$$

This is a purely mathematical approach, i.e. a formal description of a complicated experimental system which can serve as a guideline and control for the development of a sequence of detailed chemical reaction steps.

4. The oxidation of carbon monoxide

The catalytic oxidation of carbon monoxide on a noble metal like palladium follows a Langmuir-Hinshelwood-mechanism. i.e. oxygen and carbon monoxide both react in the adsorbed state [1]. A typical dependence of the rate of carbon dioxide production on the actual concentration of carbon monoxide in the reactor is depicted in Figure 4. The concentration of oxygen was kept constant in the experiment. The relatively low reaction rate was found even to decrease with increasing temperature of the reactor for a catalyst oxidized in streaming air at 330°C before use. The observed steady state of the reaction is an unstable one for the pretreated catalyst and a typical transient to higher activity and to limit cycle behavior can be seen in Figure 5.

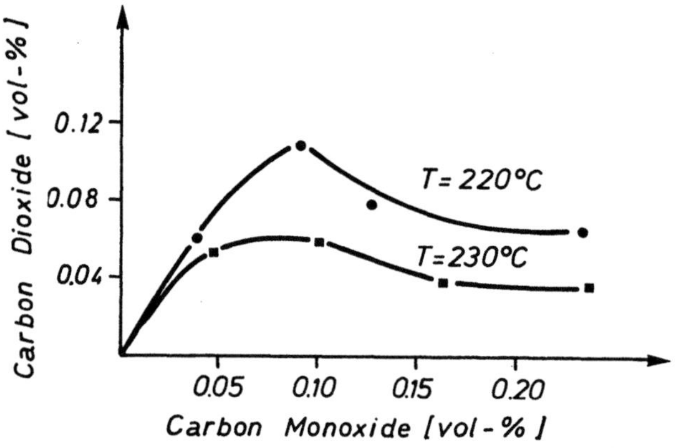

Figure 4: Dependence of the reaction rate on CO concentration in the reactor.

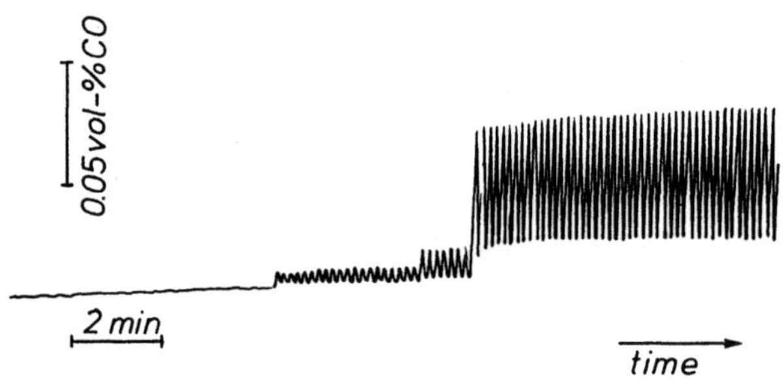

Figure 5: Transient from steady state to limit cycle behavior at 200°C. CO concentration in the feed: 0,2 vol-%.

The transient corresponds to a forming process leading to a defined condition of the catalyst. The driving mechanism of the observed oscillations is assumed to be a periodic oxidation and reduction of the palladium surface, i.e. a periodic change of the catalyst between a state of lower and higher activity [10]. The explanation is valid for a single catalytic center like a palladium crystal a few nanometers in diameter. The observation of oscillations on a macroscopic scale requires the coupling and synchronization of a large number of such centers which are spatially separated by the ceramic support. A variation in the partial pressure of CO, i.e. a concentration wave travelling across the catalyst charge is considered to be the dominant synchronizing force following arguments presented by Schwartz and Schmidt [15].

This hypothesis regarding the coupling among billions of catalytic centers can be supported by results obtained from an experimental investigation of the synchronizing effect of an external periodic variation of the CO partial pressure on autonomous oscillations [9]. The experiments were carried out in a high conversion state of the reaction from which typical short temporal breakdowns of conversion evolve. Figure 6a depicts a typical time series for 0,5 vol-% CO in the feed. The conversion is found to drop to about 75% during the temporal breakdowns from a maximum value of about 90%. The linear Fourier transform spectrum of the oscillation is shown in Figure 6b. An aperiodic character of the oscillations can be demonstrated by the calculation of a positive Lyapunov exponent [9].

Figure 7a shows the effect of forcing the autonomous oscillator (Figure 6) with an amplitude of 0,075 vol-% CO and a forcing frequency f_a = 4 min^{-1}. The corresponding linear Fourier transform spectrum (Figure 7b) shows a drastic change compared to Figure 6b, namely an entrained or phase locked state of the periodic temporal breakdown of the CO_2 production with an entrainment frequency fe = 1/2 fa.

In the case of the oscillating CO oxidation a mathematical model has been developed based upon detailed assumptions with regard to local chemical kinetics and a possible synchronizing mechanism. Gerhard and Schuster [16] proposed a 2-dimensional cellular automaton consisting of cells which move in discrete time steps through a sequence of states. After reaching a passive state the cell is reinjected into the active state at the beginning of the sequence. The cells represent catalytic centers moving between a low and a high activity state. It can be shown that an interaction by diffusive processes between neighboring cells (von Neumann or Moore-neighborhood)

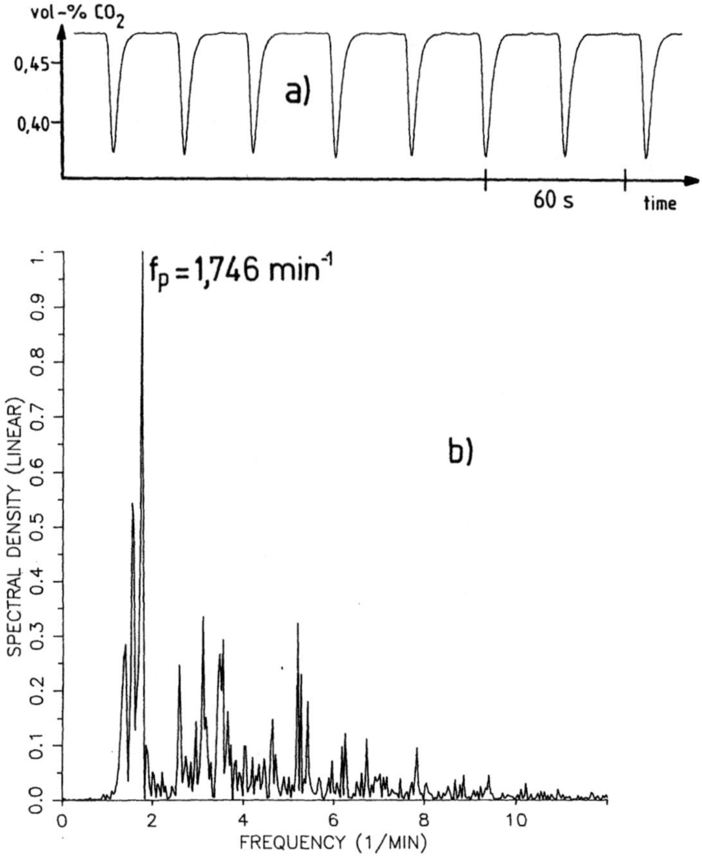

Figure 6: Time series (a) and corresponding Fourier transform spectrum (b) of the autonomous oscillation during the catalytic oxidation of CO. CO feed concentration 0,5 vol-%.

can lead to temporal order of the kind observed in the experiments. An interesting feature of the automaton in an oscillating state is the evolution of spatial order on the array of cells, a phenomenon which has also been observed by in situ measurements of the temperature distribution across a catalyst wafer during an oscillating oxidation reaction [17].

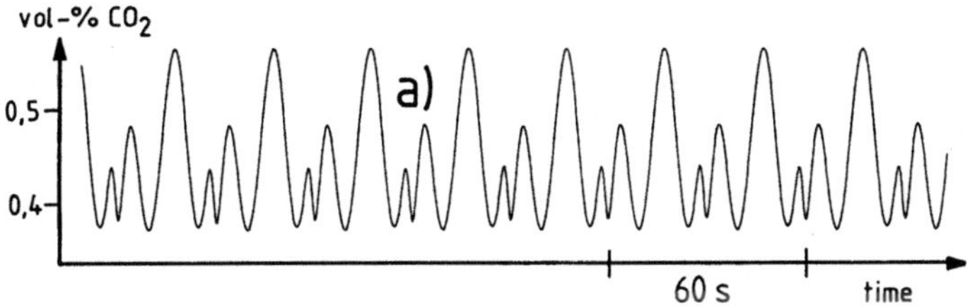

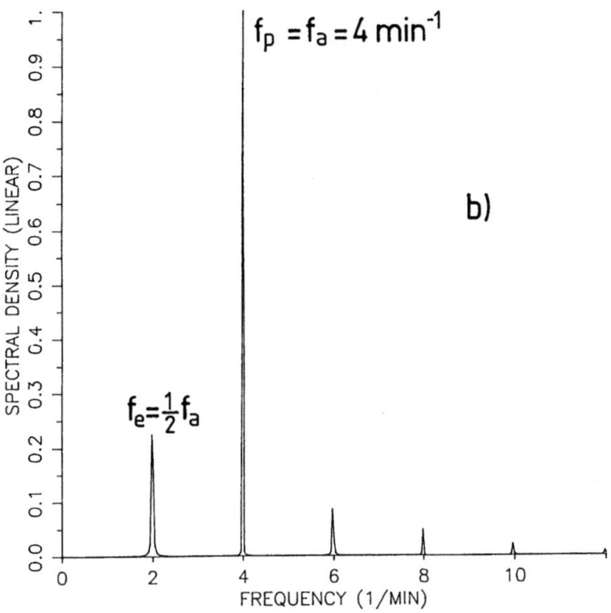

Figure 7: Time series (a) and corresponding Fourier transform spectrum (b) of the entrained oscillation. CO feed concentration 0,5 vol-%, forcing amplitude 0,075 vol-%, forcing frequency fa = 4 min^{-1}.

5. Conclusion

The experimental study, the analysis and the mathematical modelling of complicated dynamic phenomena in heterogeneous catalysis may well lead to new insights into the mechanisms of catalytic reactions. It was the aim of this article to introduce the reader into this field by presenting experimental results and approaches for the development of mathematical models in the case of two heterogeneously catalyzed oxidation reactions.

6. Acknowledgement

Figures 1, 2 and 3 were taken from ref. 8, Figures 4 and 5 from ref. 10.

References

[1] E.G. Schlosser, Heterogene Katalyse, Chemische Taschebücher, vol. 18, Verlag Chemie, Weinheim, 1972

[2] M. Sheintuch, R.A. Schmitz, Catal. Rev.-Sci. Eng. 15, 107 (1977)

[3] L.F. Razon, R.A. Schmitz, Catal. Rev.-Sci. Eng. 28, 89 (1986)

[4] H. Beusch, P. Fieguth, E.Wicke, Chem. Ing. Techn. 44, 445 (1972)

[5] Temporal Order, L. Rensing and N. Jaeger (Eds.), Springer Series in Synergetics, vol. 29, Springer, Berlin, 1985

[6] H. Kral, G-I-T Fachz. Lab. 11, 699 (1967)

[7] N.I. Jaeger, R. Ottensmeyer, P.J. Plath, Ber. Bunsenges. Phys. Chem. 90, 1075 (1986)

[8] N.I. Jaeger, R. Ottensmeyer, P.J. Plath, H. Engel-Herbert, Chem. Engng. Sci. in press (1990)

[9] P. Svensson, N.I. Jaeger, P.J. Plath, J. Phys. Chem. 92, 1882 (1988)

[10] M.M. Slin'ko, N.I. Jaeger, P. Svensson, J. Catal. 118, 349 (1989)

[11] M. Sheintuch, D. Luss, Chem. Engng. Sci. 40, 1653 (1985)

[12] M. Sheintuch, D. Luss, Chem. Engng. Sci. 42, 41 (1987)

[13] E.D. Gilles, G. Eigenberger, W. Ruppel, AIChE Journal 24, 912 (1978)

[14] H. Engel-Herbert, P.J. Plath, R. Ottensmeyer, Th. Schnelle, J. Kaldasch, Chem. Engng. Sci. in press (1990)

[15] S.B. Schwartz, L.D. Schmidt, Surf. Sci. 183, L269 (1987)

[16] M. Gerhardt, H. Schuster, Physica D 36, 209 (1989)

[17] J.R. Brown, G.A. D'Netto, R.A. Schmitz, ref. 5, p. 86

Modelling and Simulation of the Methanation from Carbon Monoxide-Rich Synthesis Gas

H.-G. Deppner and G. Schulz-Ekloff

Abstract - The methanation of carbon monoxide-rich synthesis gas is modelled by the analysis of the temperature profile of the polytropic reactor. The profile is measured by means of five thermocouples axially located in the reactor. The Arrhenius term of the kinetics is substituted by an algebraic exponential equation facilitating the parameter estimations and considering implicitly a temperature dependence of the Arrhenius parameters. Simulations demonstrate the validity of the model for experimental conditions far from the range of identification.

1. Introduction

The methanation of carbon monoxide at low temperatures (T < 550 K) with the aim to remove residual carbon monoxide from hydrogen used for catalytic processes has been studied intensively [1]. In general, heterogeneous two-phase models are used and the kinetic data can be matched by power rate laws or the introduction of adsorption coefficients. Multiplicity phenomena are described on the basis of heat and mass gradients in the catalyst grain [2].

The methanation of nearly stoichiometric synthesis gas was studied with respect to the production of SNG from coal [3] or the transport of heat via gases of high energy content [4]. Usually, adsorption coefficients are applied to fit the kinetics [5].

In the present paper the kinetics of the methanation from carbon monoxide-rich synthesis gas in a polytropic fixed bed reactor are modelled by an algebraic exponential equation. The parameters are estimated from the heat balances identified by means of five axially located thermocouples.

2. Experimental

The experimental equipment including a computer coupled (Krupp-Atlas Electronic EPR 1300) methanation reactor is described elsewhere [6]. A catalyst G-56 H (Süd-Chemie, München) containing about 15 wt.% nickel was used in pelletized (2 mm) form as a tubular (10 cm length, 2 cm diameter) fixed bed. The methanation was studied using synthesis gas of the mole ratio $H_2/CO = 4$ and varying the total pressure in the range 0.26 – 0.36 MPa and the total flux in the range 650 – 930 lh^{-1} (NTP). The gas inlet temperature was controlled to about 700 K. The temperature profile of the polytropic reactor was measured by five thermocouples located in the catalyst bed at the axial distance from the inlet of 0.5, 1.5, 2.5, 5.5 and 9.5 cm and at a radial distance from the reactor wall of about 1 cm. The model identification was carried out using the stationary temperature profiles abtained after about 1h time-on-stream.

3. Process Model

The heterogeneously catalyzed process is described by a stationary, pseudohomogeneous, one-dimensional plug flow model. The simplifications pre-assumed by this choice are justified due to the fact that the type of the temperature profiles of the reactor is dominated by the kinetics of the strongly exothermic reaction. The efforts of the model refinements are, firstly, focussed on the boundary for the heat balance, i.e. the reactor wall. The heat balance for the above mentioned model

$$(1)\ \delta(T c_p \rho \dot{v})/\delta z = \alpha_w \; 2\pi R_R (T_w - T) + Q_R \; r(-\Delta H_R\{T\})$$

was completed by a function for the temperature of the wall

$$(2)\ \delta T_w/\delta z = a_K(T - T_w) - l_a(\delta T_w/\delta z) - T_{ab}$$

$a_K(T - T_w)$ describes the heat transfer from the reactor bed to the wall, $l_a(\delta T_w/\delta z)$ the axial heat conductivity of the wall and T_{ab} the heat transfer from the wall to the environment.

Secondly, an adequate form for the kinetics ist set up. It is made use of the fact that the methanation reaction is reversible and not far from equilibrium with respect to the larger part of the reactor. In this case the reaction rate can be described by the distances from the equilibrium, i.e. by a first order relation for the equivalent rate of methane formation [7a]

$$(3)\ r(CH_4) = k\,(X(CH_4)_{eq} - X(CH_4))$$

Furthermore, if use is made of the steady-state approximation theory for heterogeneously catalyzed reactions [7b], then equation (3) might be valid for larger distances from the equilibrium, too. For the representation of the progress of the reaction rate constant k on the temperature an alternative expression instead of the usually applied Arrhenius equation is set up. The values for the terms of the Arrhenius equation differ by 8 orders of magnitude complicating the iterative estimation of the coefficients. Therefore, an algebraic exponential equation, which can be given in the most general form as differential only

(4) $dk/dT = A\ T^{B}\ e^{CT};\ C < 0,\ B > 0$

is used. The analytically integrated form

(5) $k = A/T^{3}_{K}[2+\exp\{-(T^{*}/T_{K})[-(T^{*}/T_{K})^{2}-2(T^{*}/T_{K})-2]\}]$

obtained from the values $B = 2$ and $C = -1/T_{K}$ is applied for the parameter estimations and the simulations.

4. Parameter Estimation

The steady-state process is represented by numerical integration on the axial coordinate of the reactor bed taking intervals $\Delta z \leq 1$ mm and applying Euler's method. The progress of the parameters mole fraction $X(CH_4)$, gas temperature T and wall temperature T_w is calculated using besides

(6) $T_{z+1}=T_z+(a_2\ r_{z+1}\ \Delta H_{Rz+1}+a_w 2\pi\ R_R(T_{wz}-T_z)/Q_R)(\Delta z/(\dot{V}_{z+1}\ c_{pz+1}\ \rho_{z+1}))$

and

(7) $T_{wz+1}=T_{wz} + [a_K(T_z-T_{wz})-l_a(T_{wz}-T_{wz-1})/\Delta z-T_{ab}]\Delta z$

The values $r(CH_4)$, $\dot{v}$, $\Delta H_R(T)$, c_p and ρ are actualized iteratively for each integration interval. The unknown parameters T_K, a_1, a_2, a_w, a_K, l_a, T_{wo} and T_{ab} were estimated using the gradient method of Marquardt [8]. The least squares estimation is used as criterion for the quality of the iteratively improved results. The coefficients T_K, a_2, a_w and l_a, which are invariant to the experimental conditions are kept constant for the best fit evaluation of a_1, T_{wo}, T_{ab} and a_K for a special set of experimental conditions. The evaluated coefficients are applied for the simulation experiments. The programs for the parameter estimation and the simulation, having largely identical subroutines, are written in FORTRAN 77.

5. Results and Discussion

The simulated and measured temperature profiles of the reactor bed and the estimated temperature profile of the reactor wall are given in Figure 1. The

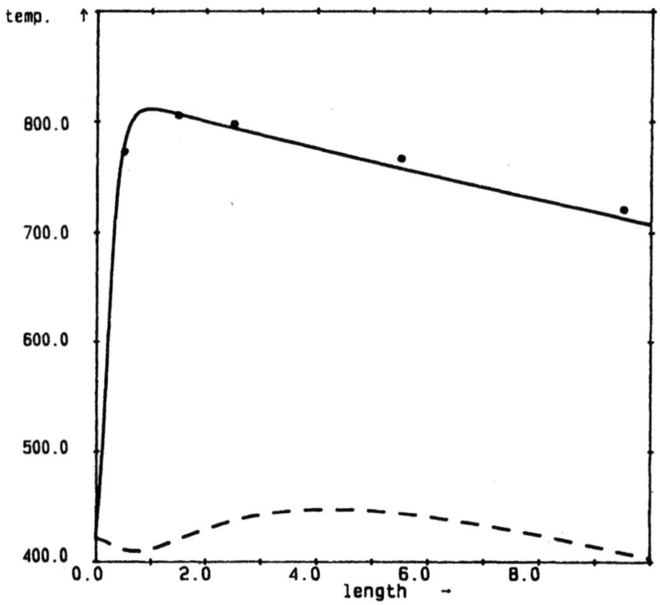

Fig. 1: Simulated (----) and measured (o) temperature profiles ($\dot{v}$ = 928.5 lh^{-1}; P = 0.26 MPa); estimated (- - -) profile of the reactor wall.

restricted range of validity for kinetics of the methanation reaction not considering the reversibility or steady-state approximation is demonstrated in Figure 2.

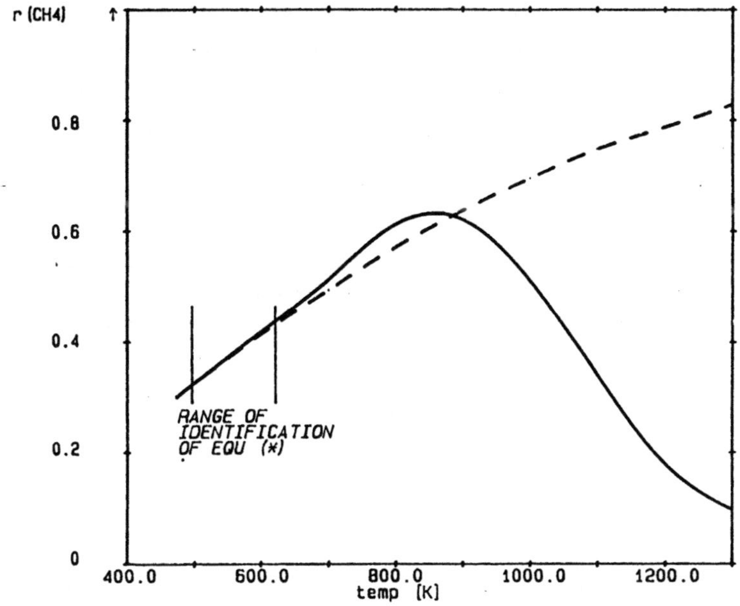

Fig. 2: Extrapolations of different functions of r (normalized values). Examples from the literature ([1], eq. 6.4, (*)) by the broken line and from eq. (3) and (5) of this paper by the full line.

Variations of the experimental parameters educt ratio, total flux and total pressure could be simulated with an accuracy of 10 K as average deviation. A drastic variation of the experimental conditions could be achieved by applying the method of catalyst dilution [9]. The larger deviations between the simulation based on the coefficients obtained for the undiluted catalyst and the experiment (Fig. 3) indicate the tendency to smooth out the expected sharp temperature step. This result indicates that the process model might be improved by considering axial heat conductivity in the catalyst bed.

The estimated function for k (eq. (5)) can be compared to the Arrhenius function yielding the temperature dependent formal activation energies given in Figure 4, which are compared with the temperature dependences of the enthalpies and free enthalpies for the methanation reaction. Similar

temperature dependences for the parameters of the formal kinetics and the thermodynamic functions might be considered as a support for the validity of the steady-state approximation for heterogeneously catalyzed reactions, especially, and the thermodynamic description of the kinetics, in general.

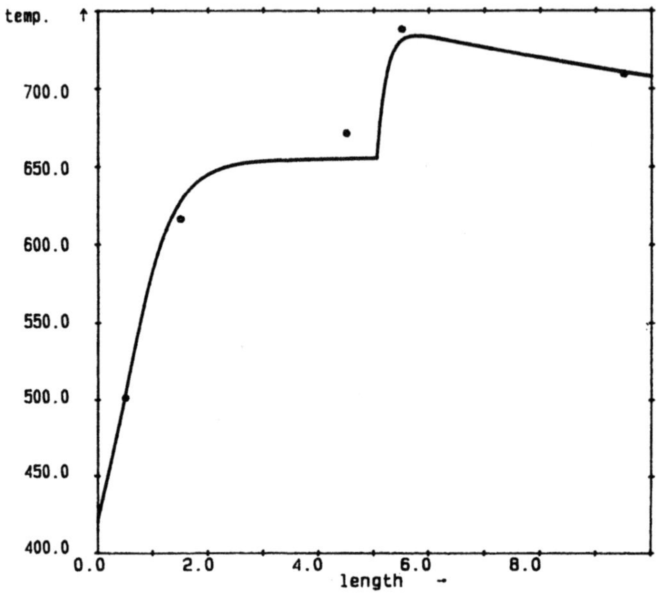

Fig. 3: Simulated (---) and measured (o) temperature profiles ($\dot{v}$ = 650 lh^{-1}; P = 0.26 MPa) for a catalyst diluted (80% alumina) at the first 5 cm of the bed.

6. Conclusions

(1) The modelling of chemical reactions can be based on the heat balances of a polytropic reactor. the model identifications make use of the temperature profiles, which can be measured easily.

(2) It is advantageous to substitute the Arrhenius term of the kinetics by an algebraic exponential equation, thus, facilitating the parameter estimations and increasing the accuracy of the results.

(3) The implicit consideration of the temperature dependence of the Arrhenius parameters, made evident from corresponding comparison of functions, gives temperature dependence of the activation energies, which can be interpreted thermodynamically.

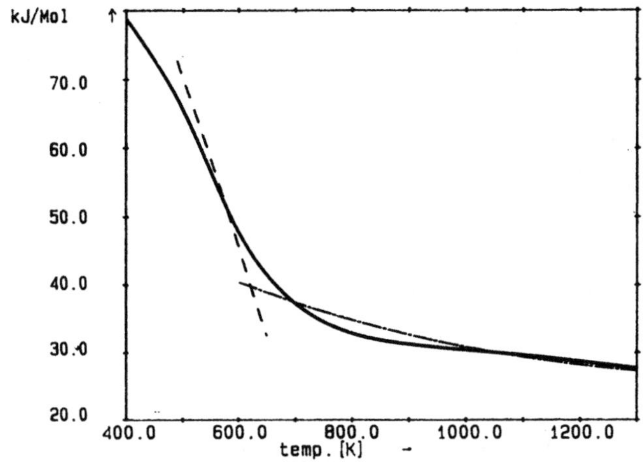

Fig. 4: Temperature dependences of the activation energy (full line) obtained by the comparison of eq. (5) to the Arrhenius type function, and the thermodynamic functions ΔG (broken line) and ΔH_R (dot and dash) for the methanation reaction.

(4) The thermodynamic interpretation of the kinetics supports the steady-state approximation theory for heterogeneously catalyzed reactions and gives an explanation for kinetic paradoxa, like the compensation effect.

List of Symbols

a_1	:	coefficient of eq. (5)	$[s^{-1}]$
a_2	:	coefficient of eq. (6)	[–]
a_w	:	coefficient of eq. (6)	$[J\ cm^{-2} \cdot s^{-1} \cdot K^{-1}]$
a_x	:	coefficient of eqs. (2) and (7)	$[cm^{-1}]$
α_w	:	heat transfer coefficient	$[J\ cm^{-1} \cdot s^{-1} \cdot K^{-1}]$
c_p	:	specific heat	$[J\ mol^{-1} K^{-1}]$
ΔH_R	:	heat of reaction	$[J \cdot mol^{-1}]$
k	:	reaction rate coefficient	$[mol \cdot s^{-1} \cdot cm^{-3}]$

l_a	:	coefficient of eq. (2)	[−]
R_R	:	reactor radius	[cm]
Q_R	:	reactor cross section	[cm^2]
r_{CH}	:	equivalent reaction rate	[$mol \cdot s^{-1} \cdot cm^{-3}$]
ρ	:	density	[$g \cdot cm^{-3}$]
T	:	temperature of the reactor bed	[K]
T^*	:	$T^* = T - 423$	[K]
T_{ab}	:	coefficient of eq. (2)	[$K \cdot cm^{-1}$]
T_K	:	coefficient of eq. (5)	[K]
T_w	:	temperature of the reactor wall	[K]
T_{wo}	:	temp. of the reactor wall at the inlet	[K]
$\dot{V}$	:	flow rate	[$cm^3 \cdot s^{-1}$]
X	:	mole fraction	[−]
z	:	axial coordinate of the reactor	[cm]

References

[1] M.A. Vannice; Catal. Rev. – Sci. Eng. 14 (1976), 153.

[2] S. Wedel and D. Luss; Ind. Eng. Chem. Fund. 23 (1984), 280.

[3] F.W. Moeller, H. Roberts and B. Britz; Hydrocarbon Processing 53 (1974), 69.

[4] H. Harms, B. Höhlein and A. Skov; Chem.-Ing.-Techn. 52 (1980), 504.

[5] H. Hammer; Erdöl und Kohle, Erdgas, Petrochemie 30 (1977), 132.

[6] D. Danziger, D. Popovic, G. Schulz-Ekloff and E. Wendland; Chem.-Ing.-Techn. 58 (1986), 48.

[7] M. Boudart; in "Kinetics of Chemical Processes", Prentice-Hall, London 1968. (a) p. 52; (b) p.65.

[8] D.W. Marquardt; J. Soc. Ind. Appl. Maths. 11 (1963), p. 431.

[9] D. Danziger, D. Popovic and G. Schulz-Ekloff; Canad. J. Chem. Eng. 61 (1983), 126.

On Parameter Estimation of Ecosystem Models

B. Breckling, A.-R. Tanha, G. Thiele

Abstract

With the background of the limitation of nature as a resource, as well as a deposit site of waste, the consideration of ecological interactions becomes more and more imperative for planning technical production processes.
Taking this into account, one of the necessary requirements is the modelling of ecological systems, the conceptual facilities and difficulties of which are investigated and discussed in this paper, with respect to an aquatic laboratory ecosystem by applying system simulation and parameter estimation.

Introduction

Because of the limited capacity of the natural environment as a resource as well as a space for deposition of wastes, the planning of industrial production processes will have to take into account its impacts on natural connexes (TRAUBOTH, 1987). The interdependencies of industrial and ecological dynamics become more and more obvious, resulting therefore in limitations for the realizability of economic projects. The problems of waste deposition (BARKOWSKI et al. 1987), of aquatic pollution (e.g. SALMONS et al.1988) and of the risks of man-induced global climatic changes (BUDYKO et al. 1988) show the consequences of mainly economically justified interference in geosphere, hydrosphere and atmosphere. The consideration of global as well as local ecological interactions is required especially if economic planning is to be based on reliable long-term predictions of its ecological consequences. The investigation of the economic options and consequences resulting from the conceptual integration of ecological effects into economic planning is the objective of recently founded institutions like the Institute for Ecological Economy Research (PFRIEM 1988).

The understanding and modelling of ecological systems as a necessary precondition for integration of ecological aspects into the planning of technical processes is conceptually difficult because the complex interaction of the dynamics of spatially distributed and self-organized biological objects in their environment has to be described in an adequate way. Even though differential and difference equation-based models are a problematical tool for a precise prediction of ecological behaviour in most of the situations, this approach can be useful for comparative studies of model performance and the observed real system behaviour. The decision about the adequacy of an assumed model structure often meets the problem that necessarily required model parameters are neither known nor directly accessible by measurement. Nevertheless, the difficulty can be circumvented in a number of cases by the application of parameter estimation techniques. This will be demonstrated in this paper for a laboratory-scale ecosystem which allows the comparative study of a system-model and the corresponding real system under controlled and well-defined conditions.

Model building and simulation of an aquatic laboratory-scale ecosystem

Parts of the biosphere, whose organismic components interact in a way that a circulation of matter is realized, are the basic units of ecosystems research. This means an ecosystem is a unit, in which the formation of organic substance from mineral components by energy uptake (primary production), their use by different organisms for secondary production (consumption) and the reduction to mineral constituents (remineralisation) takes place (ELLENBERG 1973).

Because of the external influences, which are difficult to control and the complexity of natural ecosystems, in most of the cases only parts of the ecosystem processes are investigated and modelled, without regard to the circular organisation of the complete ecosystem. To avoid this difficulty we investigated a laboratory-scale ecosystem with a reduced complexity formed by a number of well-definded, known components, so that the characteristics of such a system are accessible for investigation. We observed that the endogenous dynamics of even such a system is of the kind, that the reproduction of results is difficult and questionable.

In three parallel units we constructed an arrangement which can be taken as a maximal simplified aquatic ecosystem: the functions of primary production, consumption and remineralisation were separated in different basins and linked by a water circulation. The total volume of each system was 220 l. For primary production we selected the emerse aquatic fern *Azolla mexicana*. The basins were radiated by sun lamps. The consumers were represented by young grass carps (*Ctenopharyngodon idella*), which can grow on a pure *Azolla* diet. The function of remineralisation of fish faeces was carried out by a complex microorganism community of a sewage sludge (**Fig 1**). During the 90 days of carrying out the experiment, a well-defined quantity of system-internally grown Azolla had been transferred daily into the fish basin so that the weight of the whole *Azolla* stock could be calculated. Furthermore, a number of biotic and abiotic system-quantities were observed in adequate sampling distances. For details of the experimental design see BRECKLING and SCHOEN (1984) and MEYER (1984). Because the activities of the sub-processes in the system were observable and quantitatively determinable, we expected that the real dynamics of the system could be described by a 4-compartments model comprising the biomass of plants, of fishes, the quantities of sewage sludge and of anorganic substances. Nevertheless, actually this was not the case.

The interaction of the single components produced effects which were specific to the integration level of the whole system. The feedback within the system, generated by the limiting factor phosphate, led to changing growth conditions of the plants. Their quality as a food resource for the fishes was not constant, resulting in different growth conditions of the fishes per time unit, although they received constant food quantities. In order to model the real dynamics of even such a simplified ecosystem, a

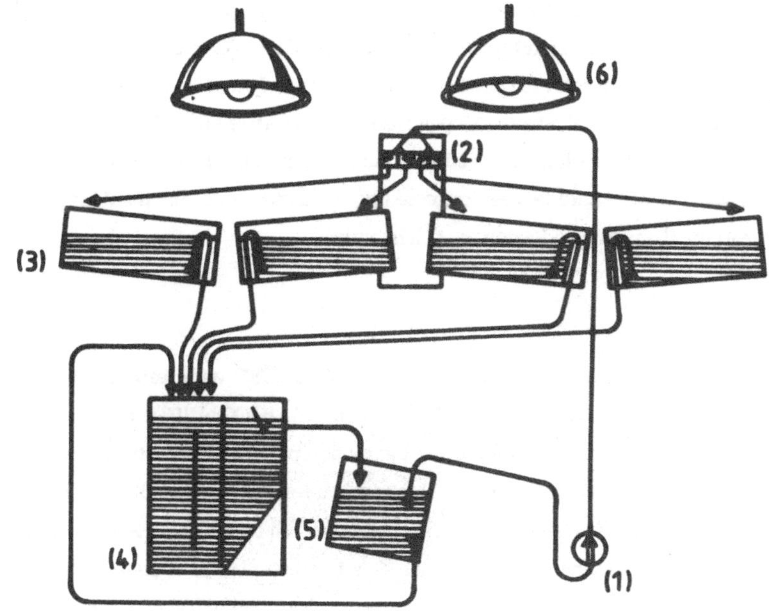

Fig 1.:
Scheme of the laboratory-scale ecosystem, illustrating the circulation of water as the means for transporting phosphate and faeces between the different basins: (1) pump, (2) water-distribution, (3) plant basins, (4) sewage treatment unit, (5) fish-basin, (6) light energy-source.

Table 1.: Denotation of estimated parameters.

AFP1	Maximum of possible phosphate uptake by *Azolla* per day
KSPA	Concentration of Phosphate in the aquatic medium, where the phosphate uptake of *Azolla* reaches half of the maximum
LPA1	Gradient of functional dependency between phosphate concentration in the medium and the phosphate uptake of *Azolla* per day
FEGE	Part of dissolved phosphate per day, which is transformed to a non plant available form due to precipitation processes
KSPI	Amount of phosphate per g plant tissue to allow *Azolla* growing with half maximum speed
PMOB	Amount of phosphate, which can be mobilized from dying parts of the plants
STST	Speed of the dying process of the plants, if the internal phosphate level of plant tissues is beyond a critical concentration
UGPA	Critical concentration of phosphate in plant tissues, below which the plants die
VERS	Amount of phosphate per g of plant tissue, where the plants virtually would stop growing (parameter fixed to 10^{-4})

model of 8 compartments turned out to be necessary (Fig 2), whose complex nonlinear interdependencies were specified by 28 parameters (BRECKLING et al. 1989). It turned out that 8 of these parameters were difficult to quantify, so assumptions had to be made. Simulation experiments showed that the results of calculations depended highly sensitive on choice of the parameter values. A number of further investigations showed that the development of the whole system with respect to the sampling frequency could be best expressed by the dynamics of the plant biomass, so that it was reasonable to use the plant production as system output.

Parameter estimation

Using the method of Output-Error Least Squares (THIELE 1987) with respect to models described by state-differential equations (RUFER 1978, TANHA and MAFTOON 1985) the possibility of matching system-measurements and simulated data by model parameter-adjustment was examined. The principle of the estimation algorithm is to compare the sum of squared differences of measured and simulated system-quantities (as error functional) and to adjust the model-parameters in an iterative way in order to minimize the error-functional. This iterative process is continued to the point where no further reduction is possible. With respect to the model-type the use of different parameter search-algorithms can be necessary.

The following provisions had to be taken into account for successful parameter-estimation:

(a) The relative adjustments applied by the search-algorithm have to be of same order for all parameters to be identified. In order to guarantee this, even if the absolute parameter values are of very different orders of magnitude, heuristically found parameter estimates $\hat{\theta}_{heu,i}$ of the order of the true parameter values θ_i have to be used enabling the estimation of transformed parameters $\theta_{t,i} := \theta_i/\hat{\theta}_{heu,i} \approx 1$ of the corresponding transformed model. Now, identical absolute adjustments $\Delta\,\hat{\theta}_{t,i}$ of transformed parameters $\hat{\theta}_{t,i}$ realize the desired identical relative adjustments $\Delta\,\hat{\theta}_i$ of original parameters $\hat{\theta}_i$.

(b) In order to simplify the parameter search it has to be limited to a relevant area in which the true parameter value must be found from biological considerations.

(c) Taking (a) into account, and with respect to the assumed true parameter area (b) identifiablity has to be proved numerically using error-free simulation data. This can be carried out by proving the existence of a finite vicinity for assumed true parameter values, the interior points of which are suitable candidates for a successful parameter search.

(d) Linear dependencies of parameters causing unidentifiabilities have to be analysed, e.g. by sensitivity analysis, and eliminated. In this sense the small output-sensitivity, with respect to parameter VERS, suggested the use of a plausible fixed value, because of its small influence on model dynamics. Furthermore, a model simplification was adequate for the elimination of linear dependence

Table 2.: Results of parameter estimation.

(1) parameter denotation, (2) heuristically determined parameter values, (3) and (5) initial values of estimation runs I and II, (4) and (6) final values of estimation runs I and II.

(1)	(2)	(3)	(4)	(5)	(6)
LPA1	0.0110	0.0032	0.00147	0.0055	0.00149
FEGE	0.3000	0.2185	0.04544	0.1500	0.04741
KSPI	0.0025	0.0081	0.01020	0.0087	0.01007
PMOB	0.2000	0.1395	0.28664	0.1000	0.27762
STST	100.00	436.68	1395.11	350.00	1261.32
UGPA	0.0140	0.0169	0.02248	0.0210	0.02228

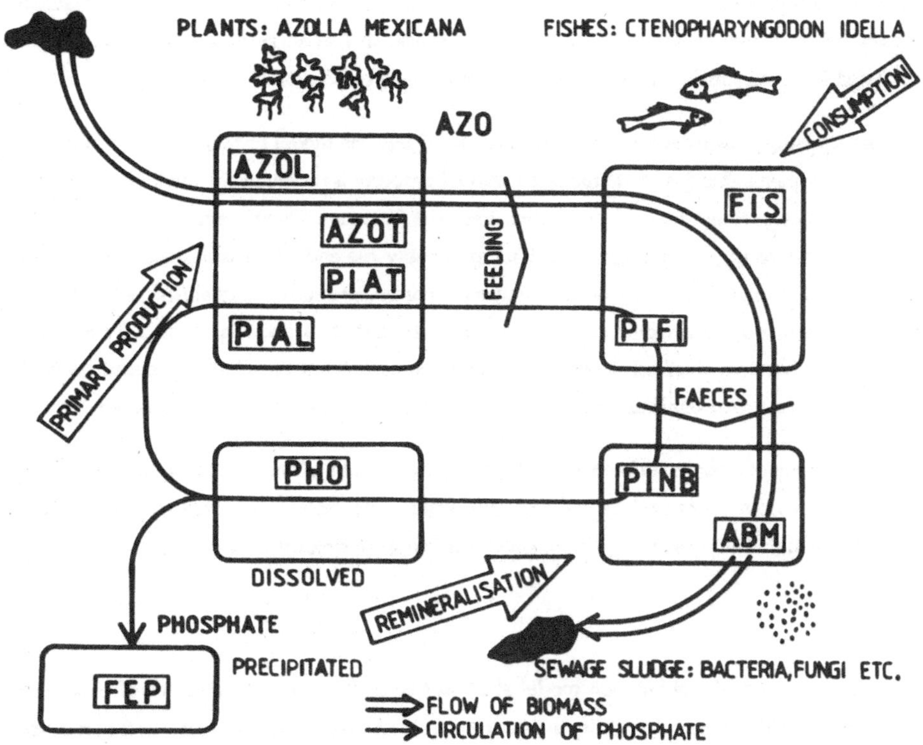

Fig 2.: Compartment structure of the simulation model.

of parameters AFP1 and KSPA, so that parameter LPA1 had to be identified instead (Denotation of the estimated parameters see in **Tab. 1**).
As a result the number of parameters to be identified could be reduced by 2.

(e) If the matching of measured system-output and simulated model-output for estimated parameters is not satisfactory, it can be conjectured that the parameter-search has stopped in a relative minimum of error-functional. In this case manual parameter correction can be tried to find an estimate of better output matching. The corresponding "corrected" estimate is often a suitable starting value for a new search-iteration resulting in a sufficient approximation of true parameters (semi-automatic parameter estimation).

Results

When estimating 8 parameters of the model in the very beginning of the experiments, different starting values resulted in significantly different estimation results. This can be caused by the existence of relative minima of error-functional. Fortunately, we found that the output matching was insufficient in these cases, even in a qualitative sense, so that these parameter estimates could be classified as not reasonably interpretable in a biological sense (**Fig. 3**).

Further investigations by sensitivity analysis showed that the model behaviour is very sensitive even to small parameter changes. A characteristic model property is its relatively small sensitivity in a first part of the measurement-interval in comparision to its final part. Therefore it is not surprising that for parameter independence to become obvious, preferably the end of measurement-interval has to be taken into account. On the basis of this analysis, 6 parameters could be qualified as identifiable with respect to Azolla-measurements as single system-output (**Tab. 2 and Fig. 4**). In addition, using the available measurements for fish biomass and for phosphate did not result in the identifiablity of more parameters, because of significantly increased estimation-error variances in this situation.

The observation of definite differences in the estimation-results of the similar experiments, studied at the same time, can be caused by the fact that system parameters, qualified as 'known', were not absolutely equal in all cases as a result of small differences in experimental conditions.

Discussion

The performance of the simulation model showed that even very small changes of parameters can cause remarkable influences on the system output. This corresponds to the fact that the sensitivities of the dynamics on parameter changes differ by orders of magnitude. Concerning the real system this means that even very small parameter changes possibly influence the system dynamics basically, for example by relatively small differences of environmental conditions. As an example a temperature difference of about 1°C in one of the parallel plants inside the climatisized laboratory caused more and

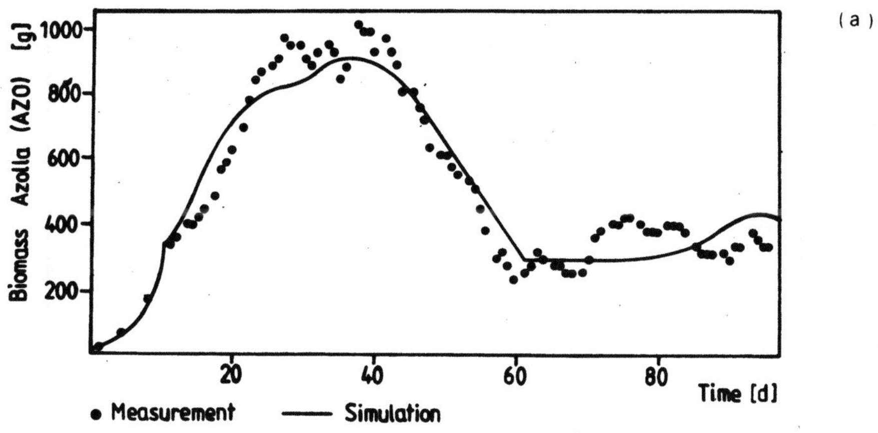

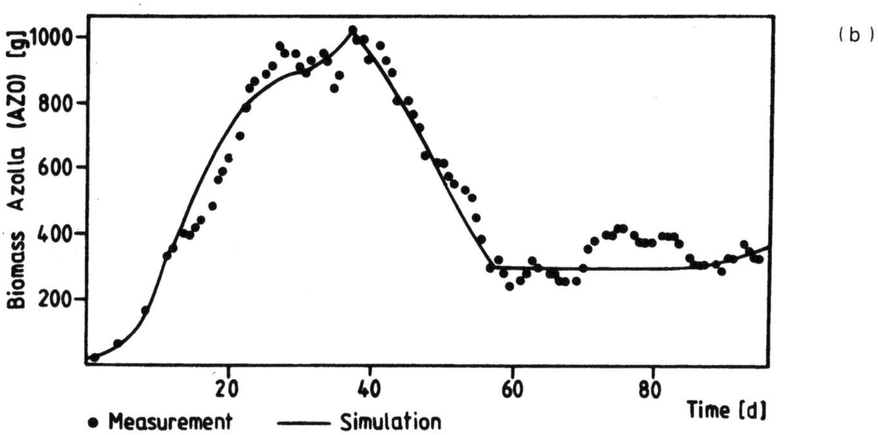

Fig 3.:
Unsatisfactory matching of simulation- and measurement-data in the case of 8 parameters to be identified: (a) matching for heuristically chosen parameters and (b) for parameters found by estimation algorithm terminating in a relative minimum of error-functional.

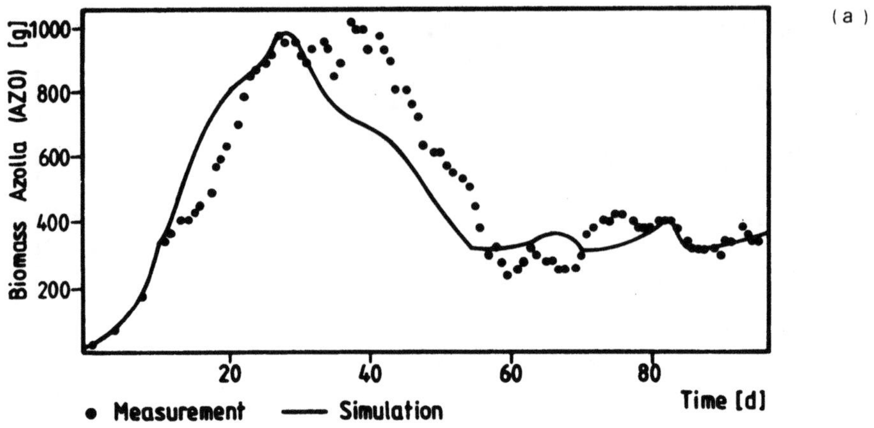

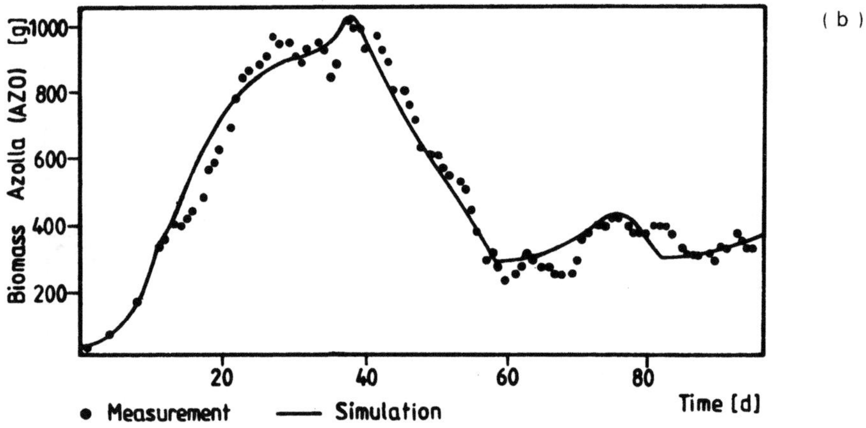

Fig 4.:
Simulated and measured behaviour of total *Azolla* biomass with (a) initial parameter values (Table 2, col. 5) and (b) estimated parameter values (Table 2, col. 6) in the case of the reduced model with 6 identifiable parameters.

more obvious differences of states, which were of the same size in the beginning. Especially in a situation like this it is difficult to decide, whether a model can not reproduce single aspects of the real dynamics for structural reasons or whether the differences are caused by invalid parameter values in a structurally valid model. Using parameter-estimation methods, we found for this model that from an ecological point of view a surprisingly good correspondence of measurements with simulation-results is achievable. This supports the assumption that the model even with regard to the sensitivity of the dynamics, can reproduce aspects of the real system in a qualitatively correct way.

Furthermore, we found that the last third of the experiment interval is important for the identifiability of parameters to be estimated. This means that the duration of the experiment can determine whether ecological parameters can be estimated successfully or not.

We found that the realized laboratory-scale ecosystem is very sensitive to changes of the conditions of the surroundings. Therefore, it is possible by parameter estimation to obtain a very precise insight into the parts of the system which are affected by these changes. As a consequence, systems of this type should be adequate for a sensitive diagnosis of changing external conditions. The investigated example of an ecosystem shows that a system, by interaction of the components, is much more sensitive than the investigation of the single components would have shown. This confirms a basic concept of ecotoxicology, which investigates the effects of toxic chemicals not only at the integration level of monospecies tests but intends obtaining more differentiated results about the effects of external impacts on ecological connexes (especially effects of chemical substances) at the level of ecological interaction in the ecosystem connex (WEIDEMANN 1989).

The sensitivity of the model supports the assumption that it will hardly be possible to use simulation models in the case of natural ecosystems for long-term prognosis about details of future developments. But especially the model-sensitivity shows an aspect, which illustrates the importance of simulation models for the understanding of effects caused by external interventions in ecological connexes: The investigation of how far it is possible to reconstruct the observed dynamics in spite of the limited knowledge about ecological interactions. Here, parameter estimation is an instrument to extend the range of cases, for which this question is decidable.

References

BARKOSWKI, D.; GÜNTHER, P.; HINZ, E.; RÖCHERT,E. (Eds.) 1987
Altlasten.
Karlsruhe (C.F. Müller)

BRECKLING, B.; SCHOEN, M. 1984
Modellierung, Simulation und Realisierung eines aquatischen Laborökosystems - eine Untersuchung zum Verhältnis von Modell und ökologischer Realität.
Diplomarbeit an der Universität Bremen, Fachbereich 2

BRECKLING, B.; TANHA, A.-R.; THIELE, G. 1989
The Simulation Model of an Aquatic Laboratory Ecosystem.
in: D.P.F. Möller (Ed.) Erwin Riesch Workshop,
System Analysis of Biomedical Processes (p. 206 - 214),
3rd Ebernburger Working Conference 7.-9. 4. 1988
Braunschweig, Wiesbaden (Vieweg)

BUDYKO, M.I.; GOLITSYN, G.; IZRAEL, Y.A. 1988
Global climatic catastrophes.
Berlin, Heidelberg, New York (Springer)

ELLENBERG, H. (Ed.) 1973
Stand und Ziele der Ökosystemforschung.
Berlin, Heidelberg, New York (Springer)

MEYER, L. 1984
Analyse von Stoffumsätzen in parallellaufenden Laboranlagen zur integrierten Pflanzen- und Fischproduktion.
Abschlußarbeit im Studiengang Bioingenieurwesen, Fachhochschule Hamburg:
Fachbereich Bio-Ingenieurwesen, Produktionstechnik und Verfahrenstechnik

PFRIEM, R. 1988
Ökologische Unternehmensführung.
Schriftenreihe des Instituts für ökologische Wirtschaftsforschung (IÖW) Nr. 13
Berlin (IÖW)

RUFER, D. 1978
User Guide for NLP:
A Subroutine Package to solve nonlinear optimisation problems.
Report-Nr. 78-07, ETH Zürich, Fachgruppe für Informatik

SALMONS, W.; BAYNE, B.C.; DUURSMA, E.K.; FÖRSTNER, U. (Eds.) 1988
Pollution on the north sea: An assessment.
Berlin, Heidelberg, New York (Springer)

TANHA, A.-R.; MAFTOON, H. 1985
Schätzung der Parameter dynamischer Systeme am Beispiel des Kardiovaskulären Systems.
Diplomarbeit an der Universität Bremen, Fachbereich 1

THIELE, G. 1987
Algorithmen zur Parameterschätzung zeitdiskreter Einfachmodelle mit dem Prozeßrechner.
VDI-Fortschrittberichte, Reihe 8, Nr. 133, VDI-Verlag

TRAUBOTH, H. 1987
Was kann die Informationstechnik für den Umweltschutz tun?
Automatisierungstechnik at 35(11) p. 431 - 442

WEIDEMANN, G. 1989
Indikation, Beurteilung und Bewertung in der Ökotoxikologie.
Mitt. Dtsch. Ges. allg. angew. Entomol. 7 (in print)

4

Process Analysis and Control

A Petri Net-Based Tool for Computer-Aided Model Building, Simulation, and Analysis of Engineering Systems

N. Bouabdallah and D. Popović

Abstract

Software tools are compulsory prerequisites for building and verifying of mathematical models of complex engineering systems. They essentially simplify the work of systems design engineer.

In the paper, a software package is presented suitable for model building, simulation, and analysis of automation systems (e.g control systems, sequential control circuitry etc). For system description Petri nets have been selected as one of most powerful mean along this line.

After the introduction of the general concept of the tool mentioned above, the functional description of its most important parts will be given.

Keywords: Petri nets, tools, computer aided design, simulation.

1. Introduction

Mathematical model building and simulation are the powerful engeneering means and efficient help to avoid operational risks and to minimize the total costs when designing control systems for plant automation. In the meantime, Petri nets have proved to be a very suitable for system model representation, especially for the representation of engineering systems[1,6,9], where they can be used for illustrative representation of static and dynamic relationships within the system, as well as for simulation an analysis of systems time behaviour[5]. In addition to this, by using Petri nets parallel processes within the system can easily be represented[7].

However, the application of Petri nets implies the use of some specific software tools the development of which is, for the time being, at its very beginning[6].

From this reason, our work aimed at development of such a tool that helps the building , analysis, and modification of Petri nets-based models of complex engineering systems. The tool developed should at the same time help to automatically check the liveness, reachability, and dead-lock freedom of the modelled system[1].

For system simulation, based on the developed model, a software simulator is avaible, one of which tunable parameter is

the time. Here, each transition should be asigned the corresponding activation time, so that the tokens, after elapsing the firing time, can be passed to the output places.
At the following stage, the general structure of the tool developed at the Institute of Automation Technology of the University of Bremen will be described, along with its main components. However, before doing so some essential questions should be answered like:

- how should individual nodes (places, transitions etc) be implemented ?
- how should the nodes be interconnected ?
- how should the "tokenplay" be executed ?
- how can the concurrent transitions simultaneously be fired ?
- how can the system be extended to the high-level Petri nets (e.g. predicate/transition nets, colored Petri nets etc.) ?

2. Basic Structure and Functionality Principle

The software tool described here was developed in programming language C on a Motorola 68000-based computer, to run under the operating system CP/M of the computer. The general software structure and the procedure of its use are represented in Fig. 1, wherefrom it is obvious that the main system

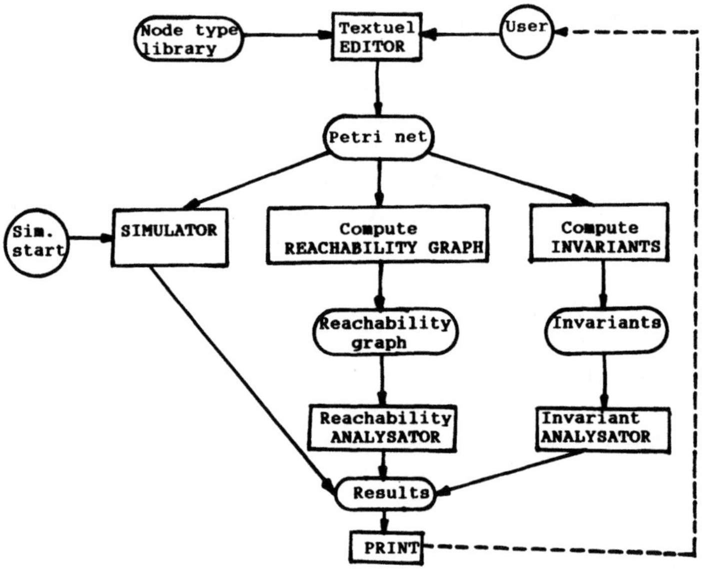

Figure 1: Basic Structure and Functioning Sequence of the Petri Net-Based Development Tool

components are:

- Library of Functions, containing the Petri net nodes like place, transition, predicate etc., implemented as corresponding procedures. Although being developed for condition/event and place/transition nets, the library can easily be extended to other Petri net types that have other node types.

- A textual Editor, designed for inputting node interconnections within the net, for automatic syntactic checks of entered data strings, and for subsequent net generation in tabular form.

- System Analysis Module, needed for computer-aided study of system behaviour, based on the use of coverability and reachability graphs and the invariants.

- Simulation Module, supported by the operating system of the computer and described in [4]. It enables to carry out the following steps in an interactive mode e.g:
  - generation and deletion of nodes and interconnections
  - display and/or modification of parameters like token number, firing time etc.
  - on-line user communication with the simulated model.

- Output and Evaluation Module.

In addition, the net nodes (place, transition etc) will also be represented by adequate software modules, and the tokens at individual places as parameters, so that the completed net parts are generated as pertaining software modules[2].

2.1 Internal Representation of Net Elements

Each net element (place, transition etc) is a process, communicating with his sequentially neighbouring elements. An element consists of some inputs, up to 4 outputs, and a block or its body (Fig. 2).
As a block input, a parameter list is used containing the source and the sink places of the respective transitions, along with the directions of token passing. Similarly, the outputs are represented by a source list of tokens leaving the block itself.

The individual blocks represent the specific procedures which determine the token passing conditions for places, check the triggering conditions of the transitions, and also trigger the transitions themselves, modelled in the net.
A net element contains the following informations:

-Name of the Element (e.g PLACE, TRANSITION)
-Address Pointer to
  -all Input labels
  -all Output labels
  -all Parameter labels like capacity, arc weights etc

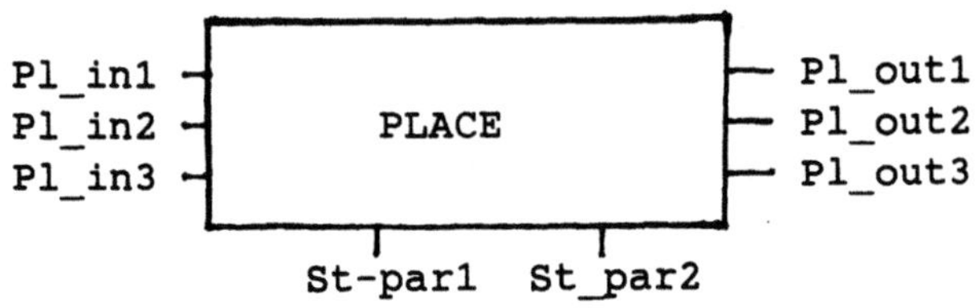

Figure 2 : Representation of a Net Element

The Informations listed above are required when generating and loading the net elements for reservation of storage needed, whereby each type of Block (or each type of procedure) will be stored once only in the Library of Functions and needs not be copied from there when initiated to run within the net. For this purpose, the reference to the function involved (via the Input/Output and Parameter Tables) will initiate it, that is used within the competence of operating system. This enables a flexible and transparent building of relatively complex net models and their simulation with a ralatively small work memory.

It is once more to be pointed out, that to each place its capacity, and actual marking is assigned as the parameter, whereby to each transition the activation time, cycle counter, and status is assigned.

2.2 Net Building Procedure

The net building procedure starts with the generation of a table of the type Net, in which all network nodes are inserted. The information transfer from one node of the net to the other one is established by reference entries stored within the Net-Table of the information sink as an address pointer to the corresponding entry of the Net-Table where the information source is stored

Net
_Address Pointer of Nodes-Table 1, _Address Pointer of Nodes_Table 2, .. _NULL

At second hierarchical level, for each net element a Node-Type Table is generated, with the information content:

NODE
-Node Label, _Address Pointer to . Node-Type, . Inputs, . Outputs, . Parameters, -NULL

It is to be noticed, that the last entry of the table is a NULL-Pointer.

This should be graphically illustrate in the following example (Fig. 3).

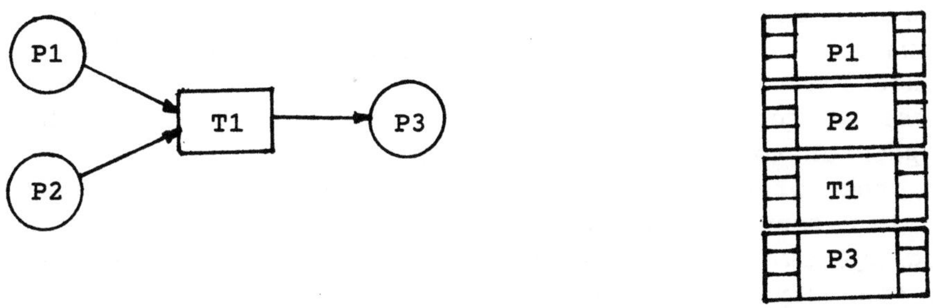

Petri Net Net Internal Representation

Figure 3: The Internal Structur of a Petri Net

Consequently, in the process of network generation the following tables will be generated and stored in the computer:

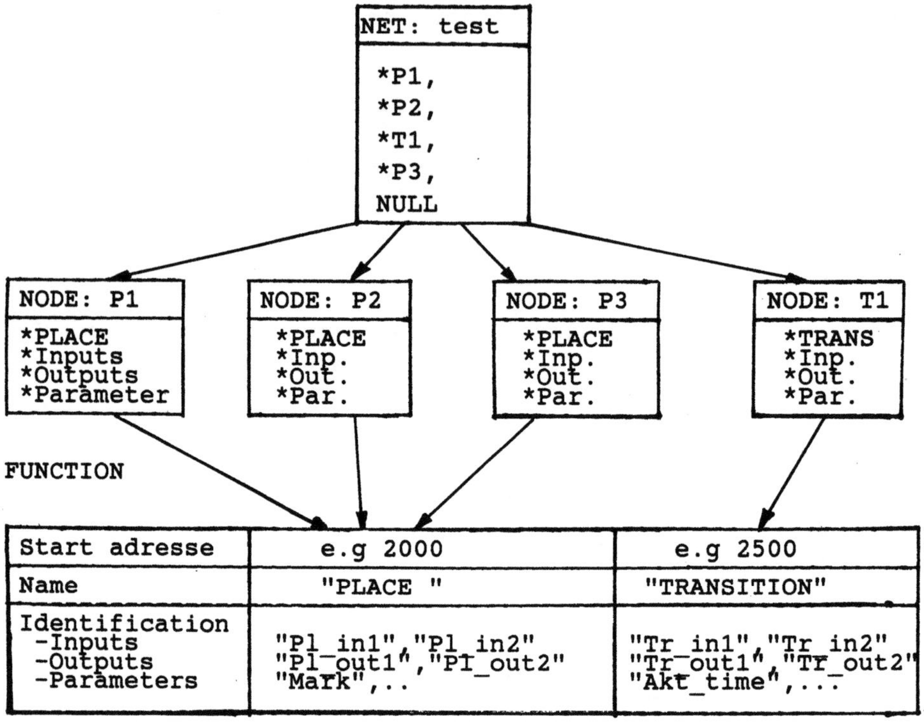

* = means that the pointer in ON

2.3 Nodes Interconnection by Address Pointers

An essential contribution described in the paper represents the software procedure that enables a direct and easy interconnection of individual net elements within the network by the use of corresponding net arcs. For this purpose, the address pointer to block input is directed to one of the outputs in the previous block[8]. The node interconnection itself, when running the model, will be established by the operating system. Here, however, one input can only be supplied by one output, whereby one output can supply several inputs (Fig. 4).

Here however care should be taken that only the different types of nodes (according to their theorical classification) can be interconnected.

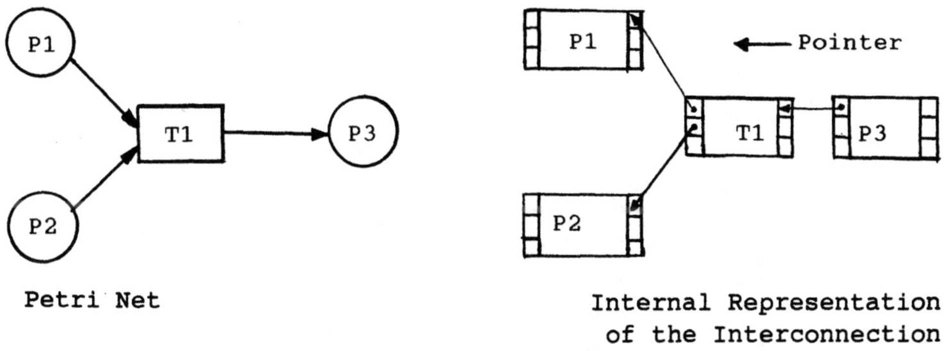

Figure 4: Symbolic and Internal Representation of the Interconnection Within a Petri Net

2.4 Incidence-Matrix Generation From the Net Structure

It is common to transform the structure of an analysis-algorithm (invariants computation and the reachability graphs) into a pertaining matrix that itself describes the network[1]. Using the internal network representation as described here, this transform can easily be carried out. This is shown in Fig. 5, where, by the use of the findout()-Procedure, it can be determined to which output a certain input of a specific block is connected. Based on such information for all nodes, the required matrix can be generated.

In order to remain within the system or memory capacity limits, the dynamic data structure is applied for internal representation of incidence matrix.

3. System Simulation: The "Token game" Principle

In the process of system simulation, all places and subsequently all transitions of a loop will repeteadly be updated. The updating of a place consists in checking the number of tokens available at its input and in storing of this number into the token counter. If the total number of tokens at the input corresponds to the associated arc weight (from place, to transition), the place outputs will be set to its TRUE-Value and the due number of tokens substracted. The number of free spaces at the place will be registreted at its outputs. In the course of transition updating, the inputs will

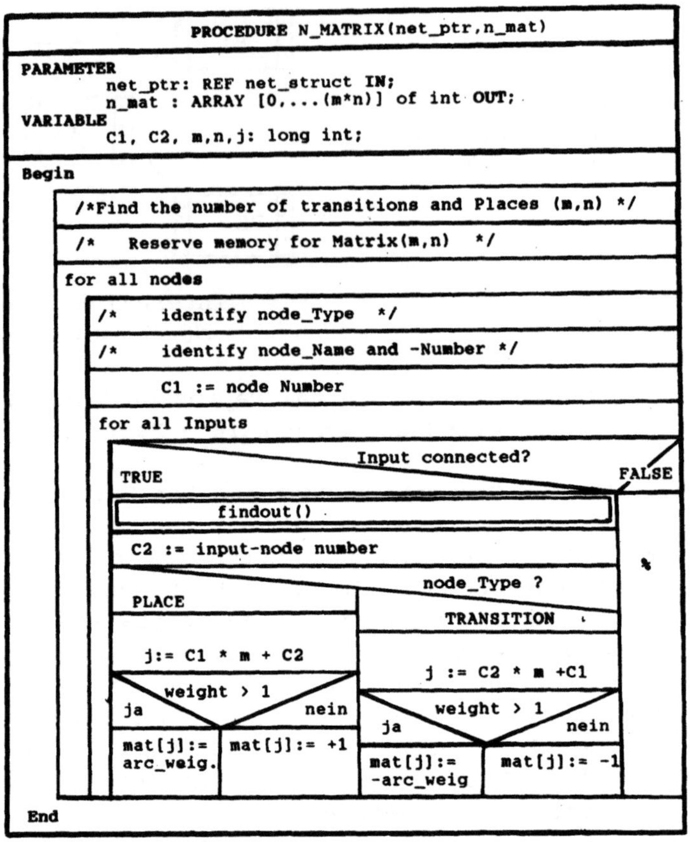

Fig. 5: Incidence-Matrix Generation Procedure

be checked on their occupancy. In the positive case, the transition will be activated, given the following conditions are met:

a) transition status is non-aktive (IDLE), Fig. 6.
b) in a conflict-case with other transitions, transition concerned must have a higher-priority level.
c) all pre-set places must be provided by the required number of places.

After activating a transition, all their inputs will be set to FALSE, the space in output places reserved, and the local time counter startet.
An activated transition will fire as soon as the activation time has elapsed. The time itself defines the time duration

requared for firing of a transition, through which the tokens will be transfered to the post-set places according to the arc weights (from transition, to place), the status switched to IDLE, release occupied places, and the clock resett to zero. Each time, before the simulation is accomplished, the system states will be updated by means of corresponding marks on the pertaining places. Furtheremore, the firing delay time (transition time-consumption) be can be defined as a stochastic parameter of a given distribution density. In this case, also a spontaneous delay-free trigering is possible.

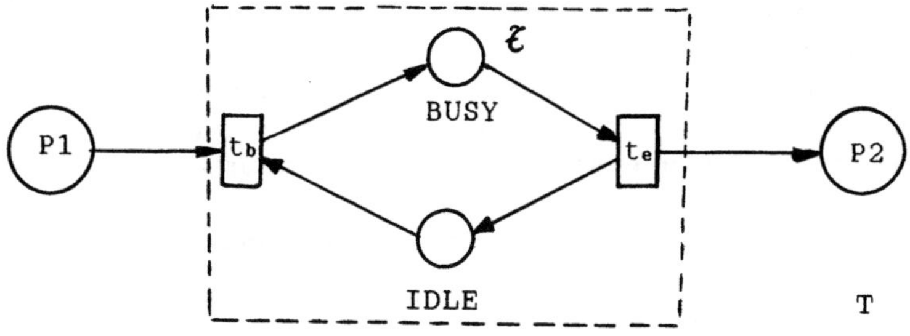

Figure 6: Zooming of a Timed Transition T

4. Summary

In the paper, the structure, funtional operability, and application of a software tool for computer-aided model building, simulation, and analysis of engineering systems was given. For model description the Petri nets have been user as a most suitable form of direct representation of static and dynamic system behaviour. The tool described is implemented on a Motorolla 68000-based computer with a CP/M operating system, on which a simulation software, written in the programming language C, can run.
The outstanding advantages of the tool are:

- -user friendliness,
- -for model building only one set of common functional modules is required,
- -analysis of model behaviour is possible,
- -extendability to higher net classes is easy to achieve,
- -both kinetic and power power performances are observable,
- -the model simulation is time-parametrized.

5. Future Actions

At present, the text oriented input for network configuration and re-configuration via corresponding commands is used. In the future, it should be replaced by a flexible graphic input facility. In addition to this, the zooming-effect on the screen will be implemented.

References

[1] Abel, D.: Modellbildung und Analyse ereignisorientierter Systeme mit Petri-Netzen. Fortschritt-Berichte VDI, Reihe 8, Düsseldorf 1987

[2] Glüer, D. und Schmidt, G.: Die Anwendung von Petri-Netzen zu Modellbildung, Simulation und Steuerungsentwurf bei flexiblen Fertigungssystemen. Automatisierungstechnik 36, Heft 12 (1988).

[3] Jensen, K.: Computer Tools for construction, modifikation and Analysis of Petri nets. Lecture Notes in Computer-Science N. 255, p.4-61, 1986.

[4] Kouwaras, M.: Entwurf und Implementierung einer Prozeßrechnerstation für die dezentrale Mehrfachregelung dynamischer Systeme. Diplomarbeit, Universität Bremen.

[5] König, R., Quäck, L.: Petri-Netze in der Steuerungstechnik VEB VerlagTechnik Berlin, 1988.

[6] Leszak, M. und Eggert, H.: Petri-Netz-Methoden und Werkzeuge. Informatik-Fachberichte 197. Springer-Verlag Berlin, Heidelberg, New-York 1989.

[7] Peterson, J.L.: Petri Net Theory and the Modeling of Systems. Verlag: Prentice-Hall, Inc., Englewood Cliffs.N. J., 1981.

[8] Popovic, D., Thiele, G., Kouwares, M., Bouabdallah, N., Wendland, E.: Conceptual design and C-implementation of a microcomputer-based programmable multiloop controller , Journal of Microcomputer Applications (1989).

[9] Reisig, W.: Petri Nets, Springer-Verlag, Berlin Heidelberg, New-York, Tokyo 1985.

Control of Distributed Parameter Systems Using Pointwise Location of Sensors and Actuators

C. Puvogel

A b s t r a c t

The principle of spatial discretization of distributed parameter systems (DPS) is shown, based on the example of diffusion processing in chemical reactors, considering a finite number of sensors and actuators. Two different design concepts of locally operating control facilities are developed, resulting in limited technical expenditure.

1. Introduction

The theory of DPS, mainly developed by mathematicians in the middle 60 's, lead to a remarkable extension of nonclassical state control. This evolution was originally induced by requirements of optimal control in chemical processing as well as electrical power technology. Progress in microelectronics during the early 70's resulted in a technological base, enabling implementation of complex identification - & control-algorithms for DPS.
Although recent integration of DPS into modern state theory lead increasingly to new application fields (medicine, biology, geology, meteorology, agricultural science, environment technology etc.), an extensive drag into industrial processing can't be observed so far. This gap might be explained by application orientated publishments on one side, rather theoretical statements on the other one.
Therefore this contribution might narrow the growing discrepancy between the advancing theory of DPS and it's missing applications in industrial processing. Consequent model reduction results in finite dimensional order, considering unavoidance of separated, locally acting sensors and actuators in technical processes.

2. State Description of DPS

DPS are described by partial differential equations. This is demonstrated using the example of space-depending, two-component's diffusion x_1, x_2 in a cylindric reactor (Fig. 1). Supposing space-dependance of both states $x_1 = x_1(z_1, z_2, t)$, $x_2 = x_2(z_1, z_2, t)$ in horizontal and radial axis for the linearized process, the system might be modelled by n = 2 coupled linear partial differential equations of the form:

$$\begin{aligned}\frac{\partial}{\partial t} x_1(z_1,z_2,t) = {} & a_{11}\frac{\partial^2}{\partial z_1^2} x_1(z_1,z_2,t) + a_{12}\frac{\partial^2}{\partial z_2^2} x_1(z_1,z_2,t) \\ & + a_{13}\frac{\partial^2}{\partial z_1^2} x_2(z_1,z_2,t) + a_{14}\frac{\partial^2}{\partial z_2^2} x_2(z_1,z_2,t) \\ & + a_{15}\, x_1(z_1,z_2,t) + a_{16}\, x_2(z_1,z_2,t)\end{aligned} \qquad (2\text{-}00)$$

It is assumed that the system's coefficients a_{ij} are none space-depending due to homogenous process environment and material coefficients.

Pointwise operating actuators are considered as distribution functions $\underline{u}_d(t)$ in equation (2-00), boundary conditions in the same way as additional operating extern inputs. Introducing spatial differential operators of the form

$$a_{zij} = a_{1ij}\frac{\partial^2}{\partial z_1^2} + a_{2ij}\frac{\partial^2}{\partial z_2^2} + a_{3ij} \qquad (2\text{-}01)$$

results in a state description

$$\frac{\partial}{\partial t}\begin{bmatrix} x_1(z_1,z_2,t) \\ x_2(z_1,z_2,t)\end{bmatrix} = \begin{bmatrix} a_{z11} & a_{z12} \\ a_{z21} & a_{z22}\end{bmatrix}\begin{bmatrix} x_1(z_1,z_2,t) \\ x_2(z_1,z_2,t)\end{bmatrix} + \begin{bmatrix} b_1 \\ b_2 \end{bmatrix} \underline{u}_\delta(t) \qquad (2\text{-}02)$$

similar to concentrated parameter systems. The system matrix is replaced in this case by an (n x n) matrix-operator $\underline{A}_z$ containing n^2 spatial operators of form (2-01).

3. Complete Discretization of DPS

The system's spatial domain is parted into n_1 sections horizontally and into n_2 sections radially, thus rendering a network of n_1 x n_2 points. Each point $z(i,j) \in z_d$ is characterized by n states of different physical unit (pressure, temperature, concentration x_1, concentration x_2 etc.) according to the system's order in time.

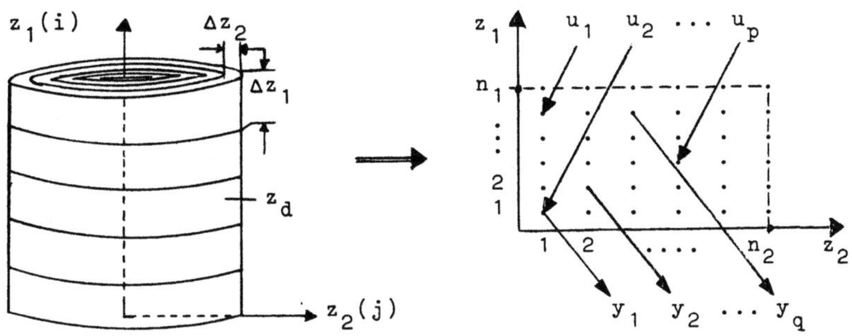

Fig. 1: Spatial (2-dimensional) dissection of DPS illustrated by discretization of cylindrical reactor/transformation to a rectangular network

Applying this discretization, the system's state vector $\underline{x}(z_1, z_2,t)$ is transformed to a finite dimensional vector $\underline{\underline{x}}(i,j,k)$ with an overall number of n x n_1 x n_2 components:

$$\underline{x}_l(i,j,k) = \begin{bmatrix} \begin{bmatrix} x_l(1,1,k) \\ \vdots \\ x_l(n_1,1,k) \end{bmatrix} \\ \vdots \quad\quad \vdots \\ \begin{bmatrix} x_l(1,n_2,k) \\ \vdots \\ x_l(n_1,n_2,k) \end{bmatrix} \end{bmatrix} \qquad l = 1,2 \tag{3-00}$$

In the case of pointwise located sensors and actuators at p respectively q places $z(i,j) \in z_d$, adequate input/output-vectors have to be defined. Indices i_o, j_o / i_1, j_1 represent special locations of pointwise operating sensors & actuators. Multiple underlining indicates the level of vector subdivision as a result of discretization.

$$\underline{u}(i_o,j_o,k) = \begin{bmatrix} u_1(i_{o1},j_{o1},k) \\ \vdots \\ u_p(i_{op},j_{op},k) \end{bmatrix} \qquad \underline{y}(i_1,j_1,k) = \begin{bmatrix} y_1(i_{11},j_{11},k) \\ \vdots \\ y_q(i_{1q},j_{1q},k) \end{bmatrix} \tag{3-01}$$

The spatially discretized model is of finite order $N = n\ n_1\ n_2$.

3.1 Algrebraization of spatial differential operators

Analogous to the introduction of time-discrete state models (replacing differential fractors by difference-equations), spatial differential operators a_{zij} (2-01) of the DPS's matrice operator $\underline{A}_z$ have to be substituted by finite dimensional submatrices $\underline{\underline{A}}_{ij}$:

$$\underline{A}_z = \begin{bmatrix} a_{z11} & a_{z12} \\ a_{z21} & a_{z22} \end{bmatrix} \xrightarrow{\text{Algebraization}} \underline{\underline{A}} = \begin{bmatrix} \underline{\underline{A}}_{11} & \underline{\underline{A}}_{12} \\ \underline{\underline{A}}_{21} & \underline{\underline{A}}_{22} \end{bmatrix} \tag{3-02}$$

In the present case, spatial differential operators a_{zij} of 2nd order have to be approximated, thus a state $\underline{x}_l(l=1...n)$ at location z(i,j) is described by

$$a_{zij}\ x_l(i,j,k) = a_3\ x_l(i,j,k)$$

$$+ \frac{a_1}{\Delta z_1^{2}} \left[x_l(i-1,j,k) - 2x_l(i,j,k) + x_l(i+1,j,k) \right] \tag{3-03}$$

$$+ \frac{a_2}{\Delta z_2^{2}} \left[x_l(i,j-1,k) - 2x_l(i,j,k) + x_l(i,j+1,k) \right] \qquad l = 1,2$$

Describing each state x_l (i,j), $i=1...n_1$, $j=1...n_2$ this way, leads to a two dimensional discrete state description:

$$\begin{aligned} \underline{\underline{x}}(i,j,k+1) &= \underline{\underline{A}}\ \underline{\underline{x}}(i,j,k) + \underline{\underline{B}}\ \underline{u}(i_o,j_o,k) \\ \underline{y}(i_1,j_1,k) &= \underline{\underline{C}}\ \underline{\underline{x}}(i,j,k) \end{aligned} \tag{3-04}$$

where submatrices $\underline{\underline{A}}_{ij}$ have a tridiagonal structure of dimension $(n_1 n_2 \times n_1 n_2)$ corresponding to spatial operators (2-01):

$$\underline{\underline{A}}_{ij} = \begin{bmatrix} \underline{A}_p & \underline{A}_k & & & 0 \\ \underline{A}_k & \underline{A}_p & \underline{A}_k & & \\ & \ddots & \ddots & \ddots & \\ & & \underline{A}_k & \underline{A}_p & \underline{A}_k \\ 0 & & & \underline{A}_k & \underline{A}_p \end{bmatrix} \qquad (3\text{-}05)$$

Submatrices $\underline{A}_p$ and $\underline{A}_k$ themselves also have diagonal/tridiagonal structure:

$$\underline{A}_p = \begin{bmatrix} p & k_1 & & & 0 \\ k_1 & p & k_1 & & \\ & \ddots & \ddots & \ddots & \\ & & k_1 & p & k_1 \\ 0 & & & k_1 & p \end{bmatrix} \qquad \underline{A}_k = \begin{bmatrix} k_2 & & & & 0 \\ & k_2 & & & \\ & & \ddots & & \\ & & & k_2 & \\ 0 & & & & k_2 \end{bmatrix} \qquad (3\text{-}06)$$

mit $p = a_3 - \frac{2a_1}{\Delta z_1{}^2} - \frac{2a_2}{\Delta z_2{}^2}$, $k_1 = \frac{2a_1}{\Delta z_1{}^2}$ sowie $k_2 = \frac{2a_2}{\Delta z_2{}^2}$

Multiple underlining of matrices indicates the level of subdivision according to the system's spatial dimension. Boundary conditions hitherto have been neglected.

3.2 Structure of discretized DPS

The given discretized system is apprehensible as a large number $(n_1 \times n_2)$ of subsystems (2nd order) corresponding via proportional factors K_1, K_2 (comp. 3-06) which might be interpreted as degree of spatial coupling. Due to the particular form of algebraized differential operators (3-03) 4 adjacent subsystems at locations $z(i\pm1,j/i,j\pm1)$ are in interaction with the subsystem at $z(i,j)$.

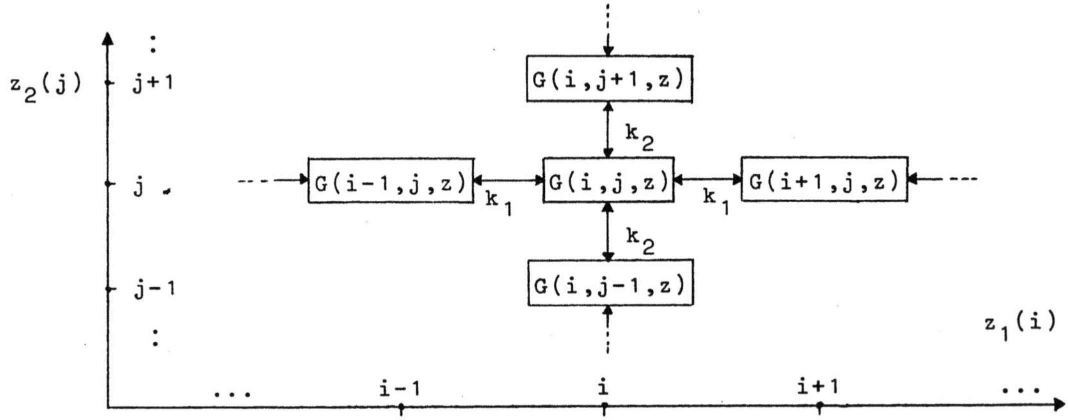

Fig 2: Structure of spatially discretized DPS/Integration of coupled subsystems, given by transfer-functions

3.3 Shape of input/output-matrices

If there is a limited number p/q of pointwise operating actuators/sensors in $z(i,j) \in z_d$, input/output-matrices in the present case of a system 2nd order have the structure:

(3-08)

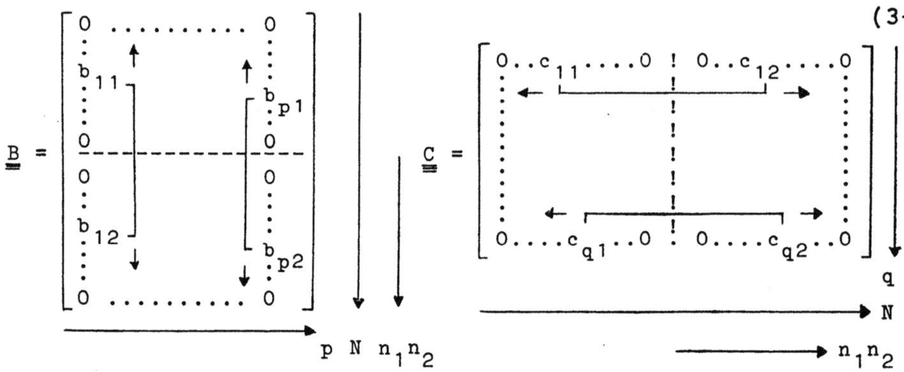

Equation (3-08) shows the influence of variation of actuator-/sensor-location in shifting elements $b_{ij}, c_{ij} \neq 0$.

3.4 Non-pointwise operating actuators

Although sensing might ensue pointwise. Actuators might act locally distributed with decreasing influence from the central point $z\ (i_o, j_o) \in z_d$ of interference. In this case the input matrice $\underline{\underline{B}}$ (3-08) contains further elements $b_{ij} \neq 0$ adjacent to the central input location.

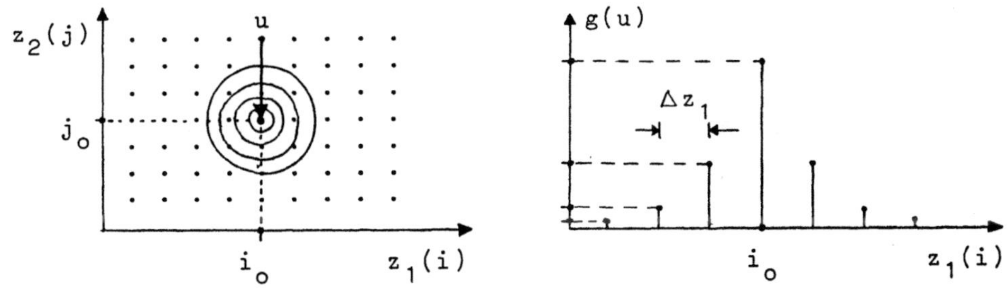

Fig. 3: Influence of local but not pointwise operating actuators with decreasing input-valuation from the central interfering location

4. System characteristics of DPS

It doesn't seem to be advisable to apply unmodified criteria of stability, controllability and observability on DPS, approximated by finite-dimensional models because:

- criteria of controllability & observability according to concentrated DPS can't be satisfied, if this is demanded for all $N = nn_1n_2$ states $x_1(i,j,k)$, using a bounded number $p,q \ll n_1n_2$ of sensors and actuators

- special gainful structure-characteristics of discretized DPS (diagonality & tridiagonality of submatrices) remain unused

- continous change of adjacent states with the same physical unit (assuming homogenous materials), allows no unpredictable large changes of state-behaviour in the environment of sensor- and actuator-location

- spatial propagation in physical & technical systems is underlying damping characteristics (k_1, k_2 <0, comp. (3-06), so that instability might be caused only at actuator locations

4.1 Introduction of local system characteristics

Introduction of local system characteristics appears to be meaningful to obtain fulfillable criteria of stability, controllability and observability.

4.1.1 Controllability

Controllability is claimed only for n states of different physical unit at q locations (i_1, j_1), considering a bounded number of pointwise operating actuators $\underline{u}(i_o, j_o, k)$ at p locations.

4.1.2 Observability

Observability is claimed only for n states of different physical unit at p locations (i_o, j_o), considering a bounded number of pointwise operating sensors $\underline{y}(i_1, j_1, k)$ at q locations.

4.2 Criteria of controllability and observability

It can be proofed /12/, that DPS are locally controllable/observable, if those states $\underline{\underline{x}}(i_1, j_1, k)$ to be controlled coincide with actuator locations $z(i_o, j_o)$, and those states $\underline{\underline{x}}(i_o, j_o, k)$ to be observed coincide with sensor locations $z(i_1, j_1)$. Additionally these $p = q$ subsystems G $(i_o = i_1, j_o = j_1, z)$ have to be individually controllable and observable in the sense of concentrated parameter systems.

4.3 Stability

Considering propagation damping characteristics in physical and technical systems (spatial coupling factors $k<1$), the problem of stability might be reduced to known criteria of concentrated parameter systems at subsystems locations $z(i_o/i_1, j_o/j_1)$.

5. Concept of local control of DPS

Considering technical expenditure and economical aspects, a spatially DPS will be controlled based on a finite number $p \ll n_1 n_2$ of locally operating actuators.

According to 4.2, controllability can't be claimed for all states $\underline{\underline{x}}(i,j,k)$ at $n_1 n_2$ locations $z(i,j) \in z_d$ if the number p of actuators at $z(i_o, j_o)$ z_d is much less $p \ll n_1 n_2$. However it is known, that in the case of DPS $(n_1 n_2 - p)$ non-controllable states of the same physical unit can't differ arbitrarily in the neighbourhood of actuator- & sensor-location.

5.1 Canonical subdivision

The discrete state description of canonically subdivided DPS (order N, comp. 3-07) with pointwise operating actuators & sensors might be represented by:

$$\begin{bmatrix} \underline{\underline{x}}_s(i_o,j_o,k+1) \\ \underline{\underline{x}}_{ns}(i,j,k+1) \end{bmatrix} = \begin{bmatrix} \underline{\underline{A}}_{11} & \underline{\underline{A}}_{12} \\ \underline{\underline{A}}_{21} & \underline{\underline{A}}_{22} \end{bmatrix} \begin{bmatrix} \underline{\underline{x}}_s(i_o,j_o,k) \\ \underline{\underline{x}}_{ns}(i,j,k) \end{bmatrix} + \begin{bmatrix} \underline{\underline{B}} & \underline{\underline{0}} \\ \underline{\underline{0}} & \underline{\underline{A}}_r \end{bmatrix} \begin{bmatrix} \underline{\underline{u}}(i_o,j_o,k) \\ \underline{\underline{x}}_r(i,j,k) \end{bmatrix}$$

Boundary conditions are considered as additional inputs according to (3-07), affecting exclusively the non-controllable subsystem. Thus the system is splitted into:

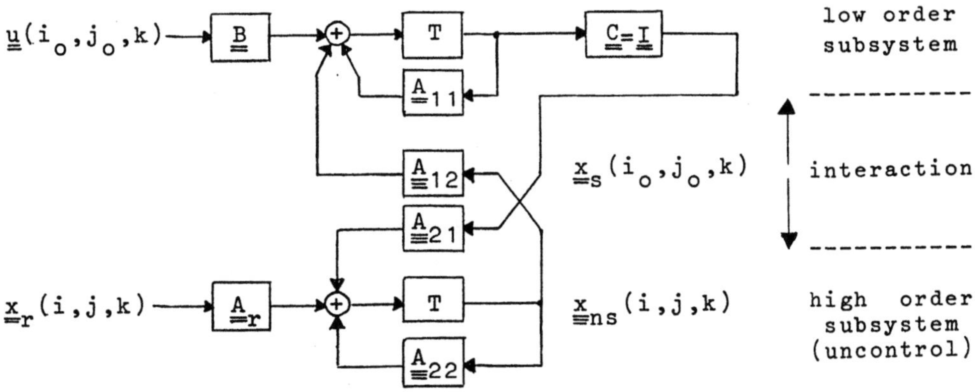

Fig. 4: Structure of a DPS in canonical subdivision

5.2 Locally operating, step optimal control of DPS in discrete state description

Due to spatial separation of sensor- & actuator location the p subsystems (order n) of the controllable state-space are decoupled: (5-01)

$$\begin{bmatrix} \underline{x}(i_{01},j_{01},k+1) \\ \vdots \\ \underline{x}(i_{op},j_{op},k+1) \end{bmatrix} = \begin{bmatrix} \underline{A}_1 & & 0 \\ & \ddots & \\ 0 & & \underline{A}_p \end{bmatrix} \begin{bmatrix} \underline{x}(i_{01},j_{01},k) \\ \vdots \\ \underline{x}(i_{op},j_{op},k) \end{bmatrix} + \begin{bmatrix} \underline{b}_1 & & 0 \\ & \ddots & \\ 0 & & \underline{b}_p \end{bmatrix} \begin{bmatrix} u(i_{01},j_{01},k) \\ \vdots \\ u(i_{op},j_{op},k) \end{bmatrix} + \underline{\underline{A}}_{12}\, \underline{\underline{x}}_{ns}(i,j,k)$$

Applying state feed back vectors $\underline{r}_\tau(t)$, $\tau = 1...p$, the system's matrices $\underline{A}_\tau$ have to be modified, such that there are nilpotent submatrices:

$$\hat{\underline{\underline{A}}}_{11}^{\,n} = \begin{bmatrix} [\underline{A}_1 - \underline{b}_1 \underline{r}_1^T]^n & & 0 \\ & \ddots & \\ 0 & & [\underline{A}_p - \underline{b}_p \underline{r}_p^T]^n \end{bmatrix} = \underline{\underline{0}} \qquad (5\text{-}02)$$

and consequently step-optimal convergence of these p locally controllable subsystems. A more detailed investigation of the subsystems (compare also 5-00): (5-03)

$$\underline{\underline{x}}_s(i_o,j_o,k+1) = \underline{\underline{A}}_{11}\underline{\underline{x}}_s(i_o,j_o,k) + \underline{\underline{A}}_{12}\underline{\underline{x}}_{ns}(i,j,k) + \underline{\underline{B}}\,\underline{u}(i_o,j_o,k)$$

shows, that there is an additional influence of non-controllable states via the coupling-matrice $\underline{\underline{A}}_{12}$, which disturbs step-optimal convergence.
This matrice $\underline{\underline{A}}_{12}$ describes the influence of adjacent, non-controllable subsystems in the immediate environment of actuator-subsystems according to the systems spatial structure (comp. Fig.2).

Introducing further sensors (or state observers) adjacent to actuator positions (comp. Fig. 6), this influence might be compensated by adding an inverse-operating matrice $\hat{\underline{\underline{A}}}_{12} = -\underline{\underline{B}}^{-1} \underline{\underline{A}}_{12}$ according to Fig. 5, thus enabling local, step-optimal behaviour

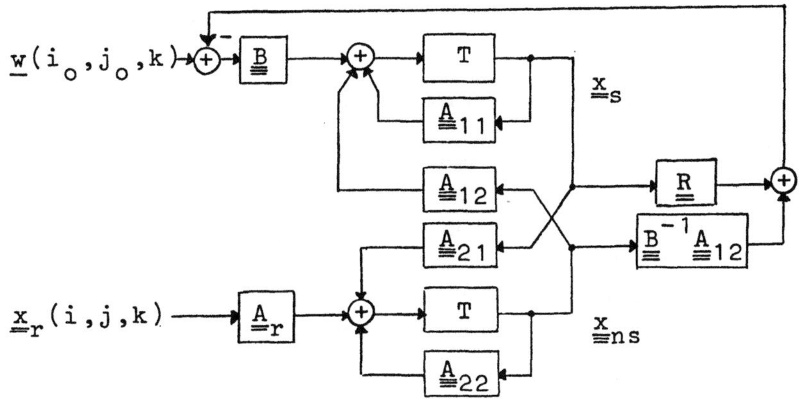

Fig.5: State controler and compensation facility

To allow inversion of input matrice $\underline{\underline{B}}$, n inputs of different physical unit $\underline{w}(i_o,j_o,k)$ have to be assumed at each actuator location, resulting in feed back matrices $\underline{\underline{R}}_\tau$, $\tau = 1...p$ instead of feed back vectors $\underline{r}_\tau^T$.
Thus complete compensation of non-controllable states in the neighbourhood of actuator-positions might be achieved.

Sensor expenditure [(np(4+1) states have to measured in the present example] might be substituted by use of state observers. In this case sensor hardware is replaced by high software expenditure, because the complete discrete model (inclusive high-order, non-controllable subsystem) has to be implemented.

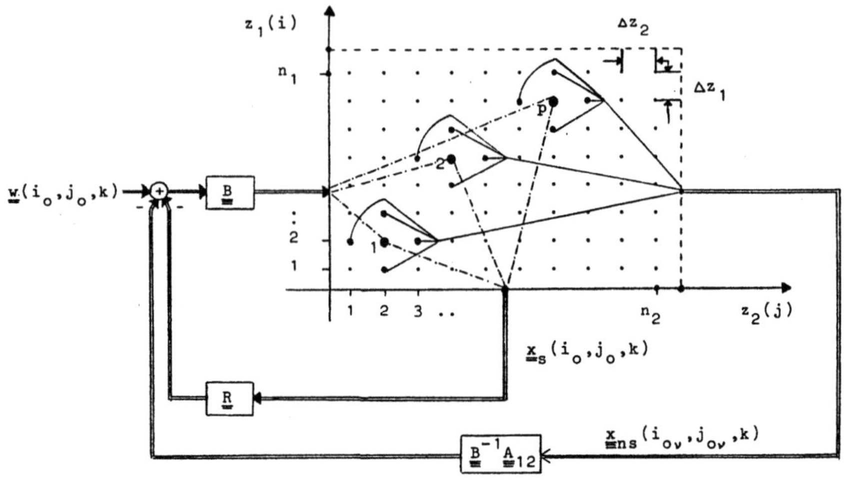

Fig. 6: Realization of local, step-optimal control, introducing additional sensors in the immediate environment of actuator positions (according to the system's spatial structure, compare Fig.2)

Above all, high actuator demand (n linear independent inputs at p locations) will restrict the suggested concept of local control.

If there is only one state of certain physical unit being influenced by adjacent, non-controllable subsystems, the compensation matrice degenerates to the form:
$\hat{\underline{\underline{A}}}_{12\tau} = - [1/b_{1\tau} \; \underline{\underline{A}}_{12\tau}]$ requiring solely single input per actuator location. If this non-controllable influence is dominantly affecting only one state, only this influence might be compensated, resulting in exponential instead of step-optimal convergence as suboptimal solution.

The suggested concept of DPS's control equals the principle of disturbance compensation.

5.3 Pointwise output-control via the DPS's boundary

In technical processing control of only one state per subsystem (order n) will be sufficient, therefore these states have to be defined as system's output.

On the other hand the system might be accessible solely via it's verge $z \in z_r$, so it has to be claimed, that those outputs to be controlled within the system's domain $z \in z_d$ don't necessarily coincide with input locations on z_r.

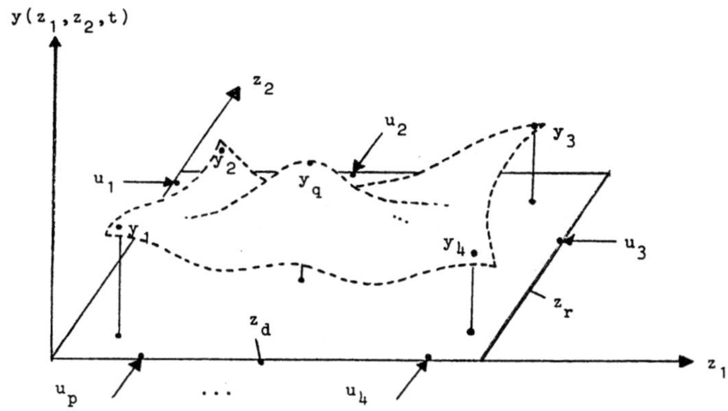

Fig. 7: Pointwise control of state distribution via the system's boundary

In this case, the system is given by it's transfer-matrice, describing the connection between p input- & q output-locations:

$$\begin{bmatrix} y_1(z) \\ \vdots \\ y_q(z) \end{bmatrix} = \begin{bmatrix} G_{11}(z) \ldots\ldots G_{1p}(z) \\ \vdots \qquad\qquad \vdots \\ G_{q1}(z) \ldots\ldots G_{qp}(z) \end{bmatrix} \begin{bmatrix} u_1(z) \\ \vdots \\ u_p(z) \end{bmatrix} \qquad (5\text{-}04)$$

Transformation of this equation leads to a decoupled form:

$$\underline{G}^* = \underline{T}\,\underline{G} \quad \text{bzw} \quad \underline{U}^* = \underline{T}\,\underline{U} \qquad (5\text{-}05)$$

such that each output y_μ is separately controlled by input u_τ by means of appropriate feed-back:

$$\begin{bmatrix} y_1(z) \\ \vdots \\ y_q(z) \end{bmatrix} = \begin{bmatrix} G_1^*(z) & & 0 \\ & \ddots & \\ 0 & & G_p^*(z) \end{bmatrix} \begin{bmatrix} u_1^*(z) \\ \vdots \\ u_p^*(z) \end{bmatrix}$$

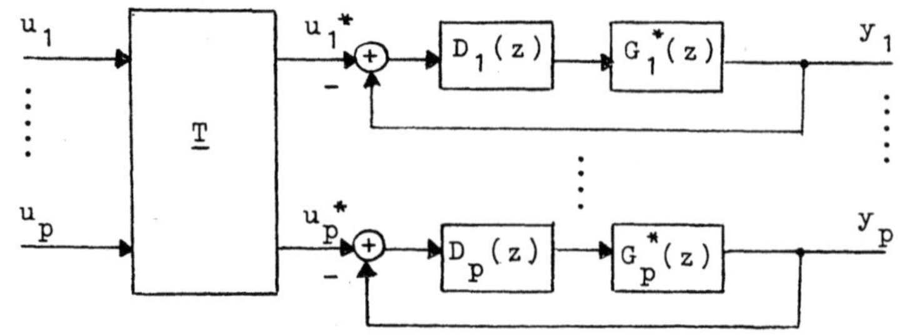

Fig. 8: Input-Decoupled MIMO-System

To avoid demand of high input energy, each output location should correspond with the closest actuator location on the system's verge z_r.
Premise of this concept is the existence of inverse matrix $\underline{G}$ (5.04), thus $p > q$ is supposed at any rate. Transfer behaviour between input u_τ at $z(i_o, j_o)$ and output $z(i_1, j_1)$ has to be approximated by a rational transfer function (order n):
$G_{\tau\mu}(z) = K^{n-1} \; [G\,(z)]^{\,no}$:
These n_o subsystems $G_{\tau\mu}$ (z) being connected by spatially coupling factors K (comp. Fig.2) represent the straight line between $z(i_o, j_o)$ and $z(i_1, j_1)$ according to Fig. 9:

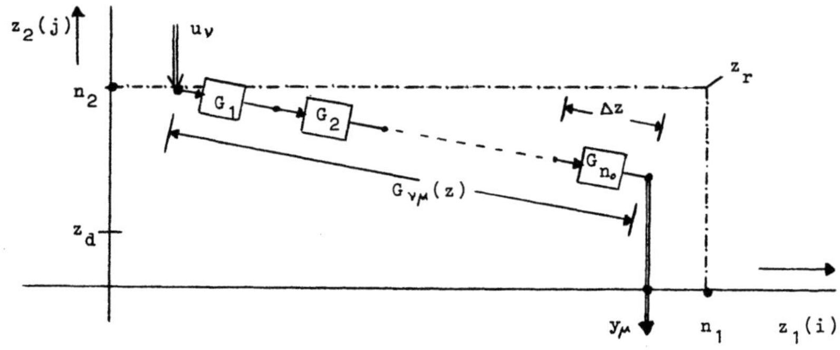

Fig. 9: Transfer model of spatially distributed system

Appendix

/ 1 / P.K.C. Wang :
Control of Distributed Parameter Systems / Advances in Control Systems ; VOL. 1; Academic Press; New York 1966

/ 2 / J.L. Lions :
Optimal Control of Systems Governed by Partial Differential Equations ; Springer Verlag ; Berlin 1971

/ 3 / A.C. Butkowskij :
Distributed Control Systems ; American Elsevier Publishing Company ; New York 1969

/ 4 / E.D. Gilles :
Systeme mit verteilten Parametern / Einführung in die Regelungstheorie ; R. Oldenbourg Verlag ; München Wien 1973

/ 5 / R.F. Curtain , A.J. Pritchard :
Infinite Dimensional Linear System Theory
Springer Verlag Berlin ; Heidelberg ; New York 1978

/ 6 / Ray , Lainiotis :
Control and System Theory , Distributed Parameter Systems Identification, Estimation & Control , Marcel Dekker Inc. 1978

/ 7 / T. Kaczorek :
Two-Dimensional Linear System ; Lecture Notes in Control & Information Sciences 75 ; Springer Verlag Berlin , New York

/ 8 / J. Klamka :
Minimum Energy Control for 2-D-Systems / Syst. Science, Vol 9

/ 9 / S.G. Tzafestas , H.J. Theodorou :
Open-Loop Deadbeat Control of Multidimensional Systems
IMACS European Meeting, Digitech. 84 / 1984 Patras Greece

/ 10 / P.K.C. Wang :
Identification in Plasma Physics
Lecture Notes in Contr. & Inform. Sciences , Band 1, Rom 1980

/ 11 / D.H. Owens :
Spatial Kinetics in Nuclear Reactor Systems in Modelling of Dynamic Systems , Peter Peregrinns , Stevenage 1980

/ 12 / C. Puvogel :
System- & Struktureigenschaften sowie Regelung örtlich verteilter Prozesse in diskreter Zustandsdarstellung
Fachberichte Elektrotechnik, Report No. 1/87 , Uni. Bremen

/ 13 / M. Köhne :
Modale Regelung des Förderrohres eines Tiefseebergbausystems
Beitrag zum VDI / VDE-Aussprachetag , Frankfurt / Main 1976

/ 14 / J.H. Seinsfeld, M. Koda :
Numerical Implementation of Distributed Parameter Filters with Application to Problems in Air Pollution
Lecture Notes in Contr. & Inform. Science , Band 1 / Rom 1976

On Modelling and Control of a Reactor for Pyrolysis

E. Koch and E. Wendland

ABSTRACT

A multivariable two-input, two-output controller is designed for control of reactor pressure and direct heating flow for warming up the pyrolysis material. In the first phase of implementation, a model with reactor pressure and feeded volume stream as controlled variables and input and output valve positions as system inputs is based on substitution of pyrolysis gases and heating gas by compressed air ignoring chemical reactions. In this way, dangers caused by explosive gas mixtures are excluded.

The controller design will be possible by standard methods for designing SISO-control loops.

1. INTRODUCTION

Increased scarcity of fossil energy resources and accelerated worsening of enviromental pollution make the pyrolysis (thermal decomposing of organic material in absence of oxigen) of outstanding interest for recycling of valuable substances.

An effective and secure process handling needs extensive knowledge about the process and considerable expense in automation technology. A laboratory-scale reactor, built in the University of Bremen, allows versatile investigations with regard to modelling and estimating of process parameters as well as to security of process handling.

For the central part of the reactor (for flow control loop and pressure control loop, coupled by the state space variable "pressure in reaction room") model equations are developed, control loops are decoupled and control algorithms are designed for decoupled control loops.

2. MODEL BUILDING

2.1 PLANT CONFIGURATION

Figure 1 gives a survey of the complete plant. The essential piece of this plant -the pyrolysis reactor- allows indirect feeding of thermic energy to pyrolysis material by 64 heat exchanging tubes as well as directly by recycling burner exhaust gas (without oxigen) or by inserting heated inert gas.

Some overpressure (100 mbar) is necessary in the reactor room to guarantee absence of oxigen. Pressure control loop secures this overpressure , which evolves during pyrolysis process by de-gasification of pyrolysis material as well as by inserting inert gas or exhaust gas for direct heating, controlled by the flow control loop.

Both control loops are coupled by the state space variable "pressure in reaction room", see Figure 2.

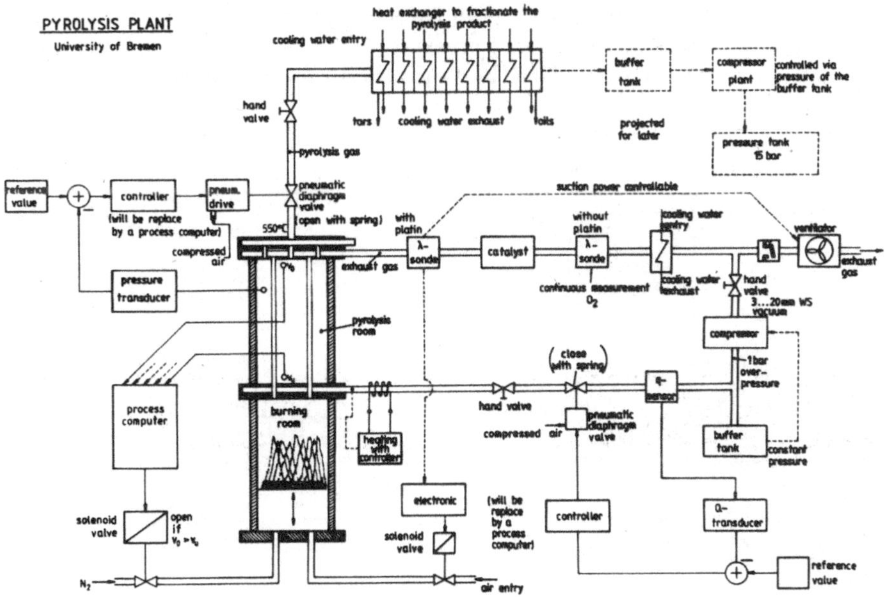

Figure 1: Pyrolysis plant at the Institut for Automation Technology of the University of Bremen

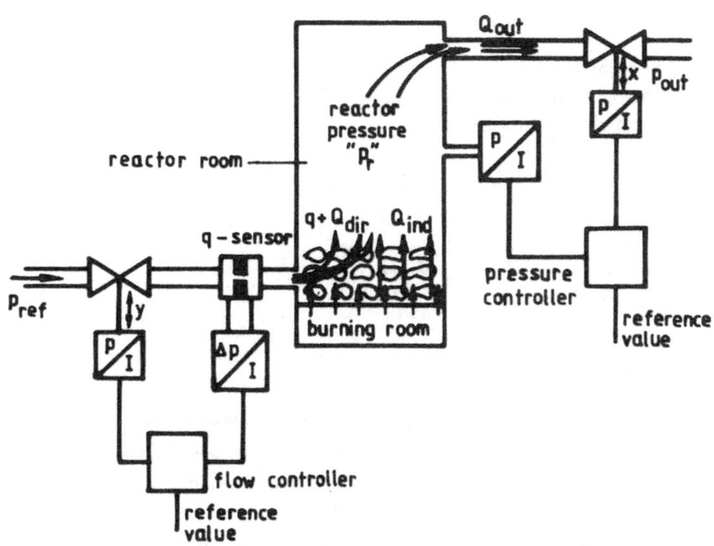

Figure 2: Simplified representation of pyrolysis reactor including pressure control loop and flow control loop

As already indicated, whole pyrolysis plant operates with compressed air for testings and simulation purposes. Increase of pressure in the reactor room caused by the pyrolysis process is imitated by additional injection of compressed air controlled via a hand valve.

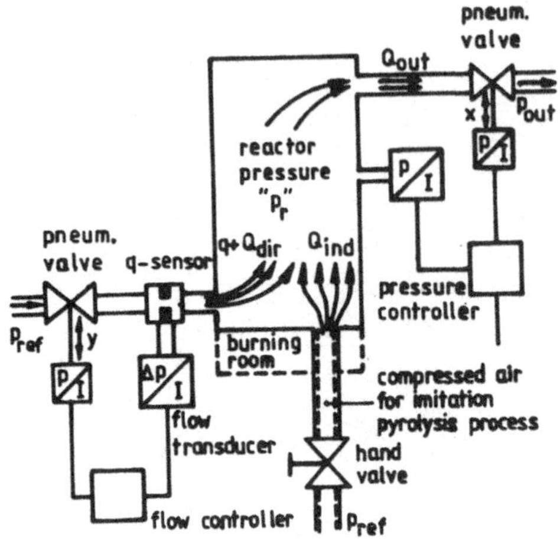

Figure 3a: Schematic representation of central part of pyrolysis reactor including pressure and flow control loop

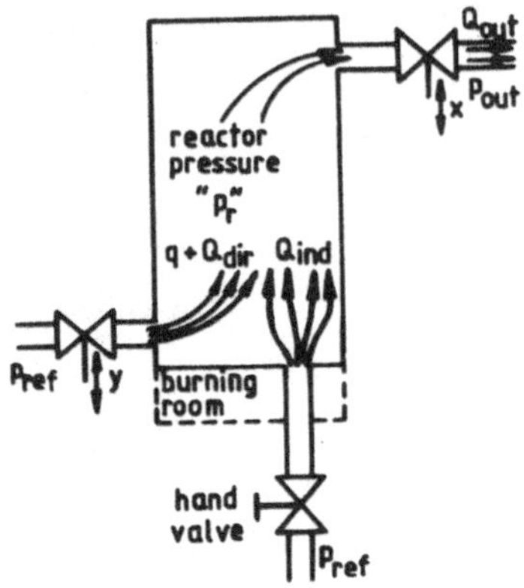

Figure 3b: Simplified representation of Figure 3a.

2.2 MATHEMATICAL MODEL

It is assumed that the hydrodynamics of the pyrolysis reactor can be discribed by the first-order nonlinear differential equation

$$\dot{p}_r = k \cdot (Q_{dir} + q - Q_{out} + Q_{ind}) \qquad (1a)$$

whereby

$$Q_{out} = A(x) \cdot \sqrt{p_r - p_{out}} \quad , \qquad (1b)$$

$$q = A(y) \cdot \sqrt{p_{ref} - p_r} \qquad (1c)$$

and

$$Q_{dir} = f(T, p_r) \qquad (1d)$$

holds.

In the equations (1a) to (1d), q is a hot gas flow for direct heating of the material to be pyrolysed. This gas is driven by the pressure p_{ref} and is controlled via position y of respective valve. Q_{dir} and Q_{ind} describe the pyrolysis yield caused by direct and indirect heating, respectively.

In the first phase of model building Q_{ind} is assumed to be constant. Furthermore, the eq. (1d) is a consequence of pyrolysis kinetics developed in [4], whereby Q_{out} is the assumed total reactor outflow at atmospheric pressure p_{out}. p_r is the reactor pressure, controlled by output valve position x, and T represents the process temperature in the reactor. Finally, k is a constant factor of suitable unit.

For small deviations of system variables around the operating point, equations (1) become

$$\Delta \dot{p}_r = k \cdot (\Delta Q_{dir} + \Delta q - \Delta Q_{out}) \qquad (2a)$$

with

$$\Delta Q_{out} = c_1 \; \Delta x + c_2 \; \Delta p_r \quad , \qquad (2b)$$

$$\Delta q = c_3 \cdot \Delta y - c_4 \cdot \Delta p_r \quad , \qquad (2c)$$

$$\Delta Q_{dir} = c_5 \cdot \Delta T + c_6 \cdot \Delta p_r \quad . \qquad (2d)$$

Assuming that the temperature deviations ΔT are proportional to thermal flow deviations Δq of direct feeded gas, e.g.

$$\Delta T = c_7 \cdot \Delta q \quad , \qquad (3)$$

from (2c) and (2d) results

$$\Delta Q_{dir} = c_5 \cdot c_7 (c_3 \cdot \Delta y - c_4 \cdot \Delta p_r) + c_6 \cdot \Delta p_r \; . \qquad (4)$$

Hence it follows, that deviation ΔQ_{dir} can be simulated along with Δq via the input valve. Assuming Q_{dir} and Q_{ind} to be constant, the following simplified model results:

$$\Delta \dot{p} = k \cdot (\Delta q - \Delta Q_{out}) \quad , \qquad (5a)$$

$$\Delta Q_{out} = c_1 \cdot \Delta x + c_2 \cdot \Delta p_r \quad , \qquad (5b)$$

$$\Delta q = c_3 \cdot \Delta y - c_4 \cdot \Delta p_r \quad . \qquad (5c)$$

3. CONTROLLER DESIGN

3.1 CONTROLLER DESIGN WITHOUT CONSIDERATION OF PHYSICAL COUPLING OF PRESSURE AND FLOW

Ignoring physical coupling of control loops controller design is based on the model, given by figures 4a and 4b , whereby flow q in pressure control loop and pressure p in flow control loop have to be interpreted as disturbances. For the pressure control loop the plant can be represented by

$$G_p = \frac{\frac{c_1}{c_2}}{\frac{1}{k \cdot c_2} \cdot s + 1} \quad . \qquad (6a)$$

Actuator S_p and sensor M_p can be discribed (according to [1] , [2]) as

$$S_p = \frac{c_{p3}}{(1 + T_{p1} \cdot s) \cdot (1 + T_{p2} \cdot s)} \quad , \qquad (6b)$$

$$M_p = \frac{c_{p4} \cdot (1 + T_{p3} \cdot s)}{(1 + T_{p4} \cdot s) \cdot (1 + T_{p5} \cdot s) \cdot (1 + T_{p6} \cdot s)} \quad , \qquad (6c)$$

whereby the following PID-structure is assumed for controller R_p:

$$R_p = \frac{c_{p5} \cdot (1 + T_{p7} \cdot s) \cdot (1 + T_{p8} \cdot s)}{s \cdot (1 + T_{p9} \cdot s)} \quad . \qquad (6d)$$

Transfer function of controller including the actuator results in

$$F_p = R_p \cdot S_p \quad . \qquad (7)$$

Both, actuator (pneumatic diaphragm valve) and measuring device (Barton transducer) are industrial standard components.

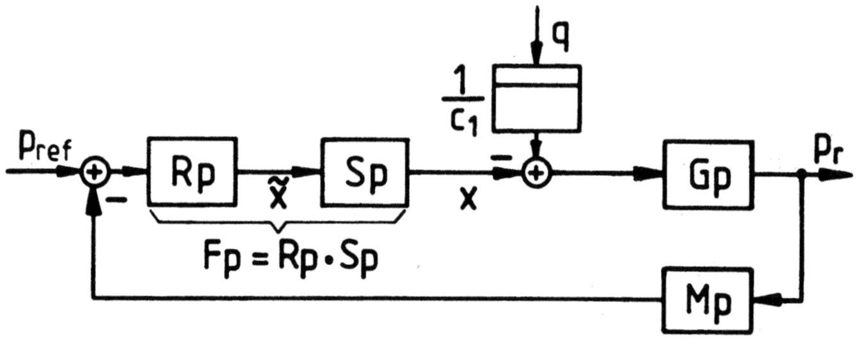

Figure 4a: Structure and notation for pressure control loop.

Actuator and sensor in flow control loop are devices of the same type as in pressure control loop. With respect to the proportional character of the flow system transfer function the controller was chosen to be of PI-structure ([1] , [2]).
Transfer functions are:

(plant) $$G_q = c_3 \quad , \qquad (8a)$$

(actuator) $$S_q = \frac{c_{q3}}{(1 + T_{q1} \cdot s) \cdot (1 + T_{q2} \cdot s)} \quad , \qquad (8b)$$

(sensor) $$M_q = \frac{c_{q4} \cdot (1 + T_{q3} \cdot s)}{(1 + T_{q4} \cdot s) \cdot (1 + T_{q5} \cdot s) \cdot (1 + T_{q6} \cdot s)} \quad , \qquad (8c)$$

(controller) $$R_q = \frac{c_{q5} \cdot (1 + T_{q7} \cdot s)}{s} \quad . \qquad (8d)$$

By analogy to F_p we have $$F_q = R_q \cdot S_q \quad (9)$$

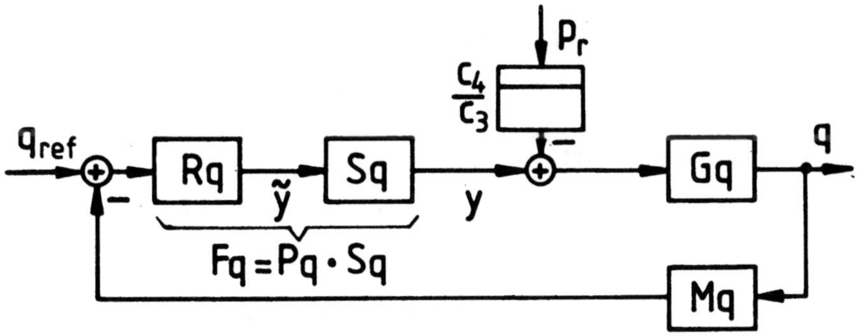

Figure 4b: Structure and notations of flow control loop

3.2 QUALITATIVE STABILITY ANALYSIS

Following analysis of stability is based on the simplified system model of coupled control loops (Fig. 5):

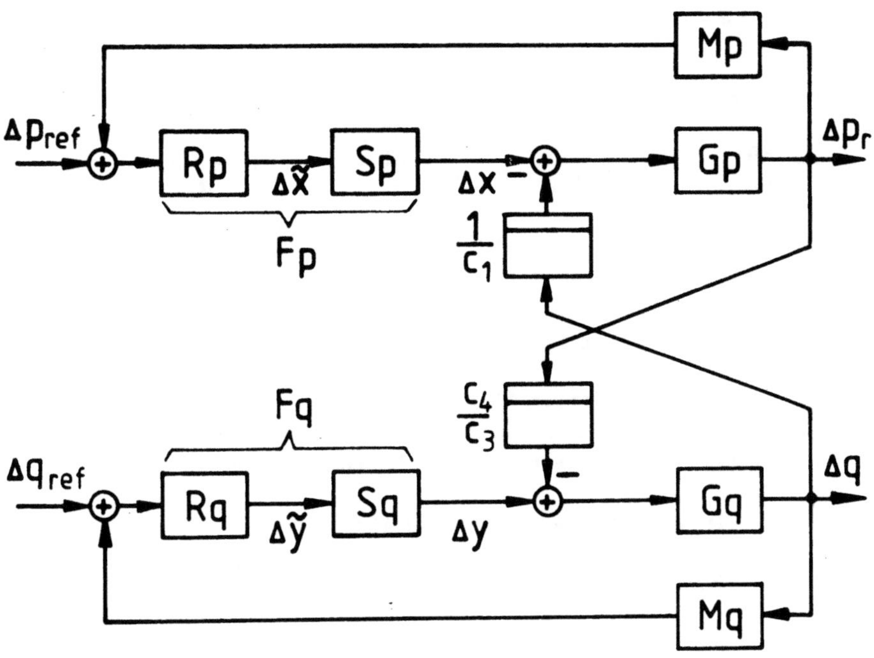

Figure 5: Complete model of coupled pressure and flow control loops

For a variation of pressure Δp_r and flow Δq in dependence of Δp_{ref} and Δq_{ref}, respectively, the following relations hold from (5) to (9):

From $$\Delta q - c_1 \cdot \Delta x = \Delta p_r \left(\frac{1}{k} \cdot s + c_2 \right) \tag{10a}$$

and $$\Delta x = F_p (\Delta p_{ref} - M_p \cdot \Delta p_r) \tag{10b}$$

follows $$\Delta p_r \left(\frac{1}{k} \cdot s + c_2 - c_1 \cdot F_p \cdot M_p \right) = \Delta q - c_1 \cdot F_p \cdot \Delta p_{ref} \tag{10c}$$

From $$\Delta q = c_3 \cdot \Delta y - c_4 \cdot \Delta p_r \tag{11a}$$

and $$\Delta y = F_q \cdot (\Delta q_{ref} - M_q \cdot \Delta q) \tag{11b}$$

follows $$\Delta q \cdot (1 + c_3 \cdot F_q \cdot M_q) = c_3 \cdot F_q \cdot \Delta q_{ref} - c_4 \cdot \Delta p_r \, . \tag{11c}$$

Elementary transformations of (10c) and (11c) result in the following equations of coupled system:

$$\Delta p_r = \frac{c_3 \cdot F_q \cdot \Delta q_{ref} - c_1 \cdot F_p \cdot \Delta p_{ref} \cdot (1 + c_3 \cdot F_q \cdot M_q)}{\left(\frac{1}{k} \cdot s + c_2 - c_1 \cdot F_p \cdot M_p \right) \cdot (1 + c_3 \cdot F_q \cdot M_q) + c_4} \tag{12a}$$

$$\Delta q = \frac{c_3 \cdot F_q \cdot \Delta q_{ref} \cdot \left(\frac{1}{k} \cdot s + c_2 - c_1 \cdot F_p \cdot M_p \right) + c_4 \cdot c_1 \cdot F_p \cdot \Delta p_{ref}}{\left(\frac{1}{k} \cdot s + c_2 - c_1 \cdot F_p \cdot M_p \right) \cdot (1 + c_3 \cdot F_q \cdot M_q) + c_4} \, . \tag{12b}$$

In this equations terms M_p, M_q, F_p and F_q are defined as:

$$M_q := \frac{Z_q}{N_q} \, , \quad M_p := \frac{Z_p}{N_p} \, , \quad F_q := \frac{\tilde{Z}_q}{\tilde{N}_q} \, , \quad F_p := \frac{\tilde{Z}_p}{\tilde{N}_p} \, . \tag{13}$$

Inserting (13) in equation (12) and multiplying this equation by $\tilde{N}_p \cdot N_p \cdot \tilde{N}_q \cdot N_q$ results in the following transfer function denominator:

$$N := \left[\left(\frac{1}{k} \cdot s + c_2 \right) \cdot N_p \cdot \tilde{N}_p - c_1 \cdot Z_p \cdot \tilde{Z}_p \right] \cdot \left(N_q \cdot \tilde{N}_q + c_3 \cdot Z_q \cdot \tilde{Z}_q \right) + c_4 \cdot \tilde{N}_p \cdot \tilde{N}_q \cdot N_p \cdot N_q \tag{14}$$

with N being a polynom in complex variable s.

By inserting transformed equations (6) and (8) in (14) the coefficients a_o and a_n of the denominator polynom $N = a_o + a_1 s + \ldots + a_n s^n$ become from

$$N_p = (1 + T_{p4} \cdot s) \cdot (1 + T_{p5} \cdot s) \cdot (1 + T_{p6} \cdot s)$$

$$\tilde{N}_p = s \cdot (1 + T_{p9} \cdot s) \cdot (1 + T_{p1} \cdot s) \cdot (1 + T_{p2} \cdot s)$$

$$Z_p = C_{p4} \cdot (1 + T_{p3} \cdot s)$$

$$\tilde{Z}_p = C_{p3} \cdot C_{p5} \cdot (1 + T_{p9} \cdot s) \cdot (1 + T_{p8} \cdot s)$$

$$N_q = (1+T_{q4}\cdot s)\cdot(1+T_{q5}\cdot s)\cdot(1+T_{q6}\cdot s)$$

$$\tilde{N}_q = s\cdot(1+T_{q1}\cdot s)\cdot(1+T_{q2}\cdot s)$$

$$Z_q = C_{q4}\cdot(1+T_{q3}\cdot s)$$

$$\tilde{Z}_q = C_{q3}\cdot C_{q5}\cdot(1+T_{q7}\cdot s) \quad :$$

$$a_o = -C_1\cdot C_3\cdot(Z_p\cdot\tilde{Z}_p\cdot Z_q\cdot\tilde{Z}_q)\Big|_{s=o}$$
$$= -C_1\cdot C_3\cdot C_{p4}\cdot C_{p3}\cdot C_{p5}\cdot C_{q4}\cdot C_{q3}\cdot C_{q5} \qquad (15a)$$

$$a_n = \frac{1}{k}\cdot T_{p4}\cdot T_{p5}\cdot T_{p6}\cdot T_{p9}\cdot T_{p1}\cdot T_{p2}\cdot T_{q4}\cdot T_{q5}\cdot T_{q6}\cdot T_{q1}\cdot T_{q2} \qquad (15b)$$

- If the constants C_{pj}, C_{qj}, T_{pj}, T_{qj} ($j \in \{1 \ldots 9\}$) and k are positive, a_o is negative and a_n is positive from (15). According to the ROUTH-criterion the control system is instable in this case.
- Due to physical reasons constants C_1, C_3, k and all T_{pj} as well as all T_{qj} are positive. Hence it follows from (15a) that the product $C_{p3}\cdot C_{p4}\cdot C_{p5}\cdot C_{q3}\cdot C_{q4}\cdot C_{q5}$ has to be negative as a necessary condition for stability of the control system.
  In the real laboratory plant this is secured by negative constant C_3 of pressure actuator and all other constants being positive.

Because controller synthesis for the multivariable system is very complex, it is an obvious alternative to decouple both control loops and to use standard methods for SISO-controller synthesis. In the following section the decoupling equations for the system are derived.

3.3 CONTROLLER DESIGN BY DECOUPLING OF PRESSURE AND FLOW CONTROL LOOP

Following model equations can be derived from (10a) and (11a) (see figure 5) :

$$\Delta p_r = \frac{\Delta q}{\frac{1}{k}\cdot s + c_2} - \frac{c_1}{\frac{1}{k}\cdot s + c_2}\cdot S_p\cdot \Delta x \qquad (16a)$$

$$\Delta q = -c_4\cdot \Delta p_r + c_3\cdot S_q\cdot \Delta y \qquad (16b)$$

Elementary transformations lead to

$$\Delta p_r = \frac{1}{\frac{1}{k}\cdot s + c_2 + c_4}\cdot (c_3\cdot S_q\cdot \Delta y - c_1\cdot S_p\cdot \Delta x) \qquad (17a)$$

$$\Delta q = \frac{1}{\frac{1}{k}\cdot s + c_2 + c_4}\cdot\{c_4\cdot c_1\cdot S_p\cdot\Delta x + c_3\cdot S_q\cdot(\tfrac{1}{k}\cdot s + c_2)\cdot\Delta y\} \quad . \qquad (17b)$$

With $$A_{11} := \frac{-c_1 \cdot S_p}{\frac{1}{k} \cdot s + c_2 + c_4} \quad ; \quad A_{12} := \frac{c_3 \cdot S_q}{\frac{1}{k} \cdot s + c_2 + c_4}$$

$$A_{21} := \frac{c_4 \cdot c_1 \cdot S_p}{\frac{1}{k} \cdot s + c_2 + c_4} \quad ; \quad A_{22} := \frac{c_3 \cdot S_q \cdot (\frac{1}{k} \cdot s + c_2)}{\frac{1}{k} \cdot s + c_2 + c_4} \qquad (18)$$

the equations (17) can be represented in blocks-diagram form by (Figure 6):

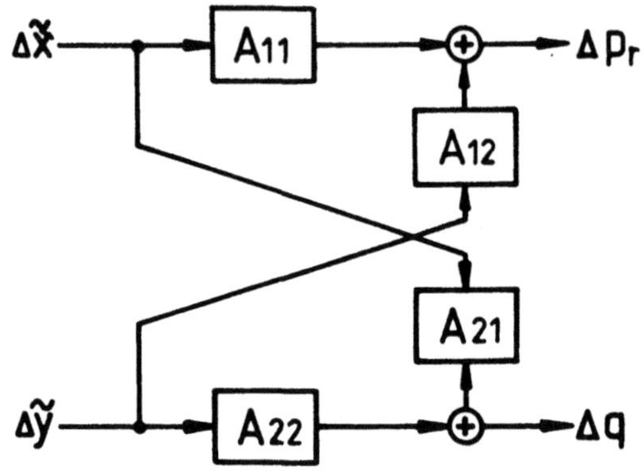

Figure 6 : Structure of plant model

In section 3.2 it was proved that ignoring the inner system couplings controller synthesis may lead to instability. In order to avoid this problem it is reasonable to place a decoupling matrix $\mathbf{B} = [\, B_{ij} \,]$ before the system matrix $\mathbf{A} = [\, A_{ij} \,]$ as shown in Fig. 7 and to select $\mathbf{B}$ so that a structure results, which is shown on right side of Fig.7.

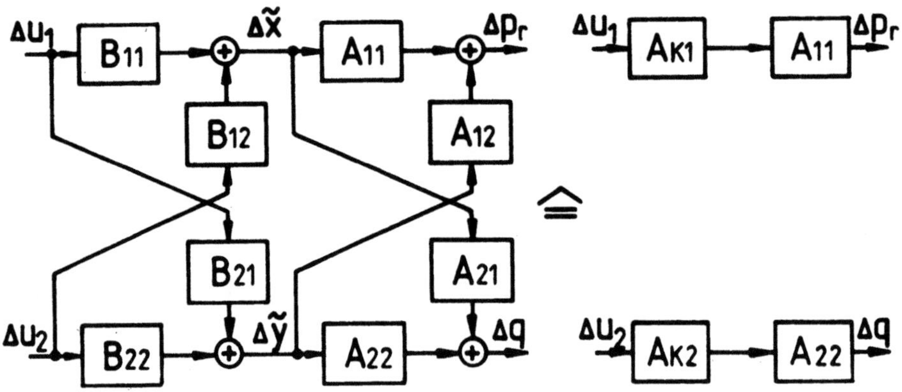

Figure 7: Decoupling of control loops

Elementary transformations yield

$$B_{11} = \frac{A_{11} \cdot A_{22} \cdot A_{K1}}{\det(A)} \quad ; \quad B_{21} = \frac{-A_{11} \cdot A_{21} \cdot A_{K1}}{\det(A)}$$

$$B_{12} = \frac{-A_{12} \cdot A_{22} \cdot A_{K2}}{\det(A)} \quad ; \quad B_{22} = \frac{A_{11} \cdot A_{22} \cdot A_{K2}}{\det(A)} \tag{19}$$

Assuming elements A_{ij} to be known and substituting for A_{ij} in equation (19) it follows

$$B_{11} = \frac{(\frac{1}{k} \cdot s + c_2) \cdot A_{K1}}{\frac{1}{k} \cdot s + c_2 + c_4} \quad ; \quad B_{21} = \frac{-c_1 \cdot S_p \cdot c_4 \cdot A_{K1}}{c_3 \cdot S_q \cdot (\frac{1}{k} \cdot s + c_2 + c_4)}$$

$$B_{12} = \frac{c_3 \cdot S_q}{c_1 \cdot S_p} \cdot \frac{(\frac{1}{k} \cdot s + c_2) \cdot A_{K2}}{\frac{1}{k} \cdot s + c_2 + c_4} \quad ; \quad B_{22} = \frac{(\frac{1}{k} \cdot s + c_2) \cdot A_{K2}}{\frac{1}{k} \cdot s + c_2 + c_4} \tag{20}$$

A_{k1} and A_{k2} represent the controller transfer functions of decoupled loops.

4. CONCLUSIONS

- By decoupling the control loops as shown in Fig. 7 it is possible to design controllers A_{k1} and A_{k2} by standard SISO synthesis methods resulting in desired behavior of reactor pressure p_r and flow q.
- Equations (19) and (20) define decoupling matrix **B** which has to be placed before real system.
- Decoupling with respect to the V-structure of plant model (see [3]) which seems to be reasonable at first glance from Fig.5 could not be selected because of difficulties in realisation due to the necessity of taking into account the transfer functions of actuators and sensors in the design.

5. REFERENCES

[1] *Busekros, L., J. Cordes, H.-W. Philippsen, H. Schröter, P. Walerius, (1983). Beschreibung einer Bartonzelle als Meßwertgeber einer Durchflußstrecke in ihrem statischen und dynamischen Verhalten. Universität Bremen, Fachber.1, AV-Bericht, Labor Prakt. Regelungstechnik*

[2] *Cordes, J., H.-W. Philippsen, H. Schröter, P. Walerius, (1985). Simulationsprogramme für lineare, zeitinvariante Mehrfachsysteme in zeitdiskreter und -kontinuierlicher Zustandsraumbeschreibung mit Anwendungen am Beispiel eines Druck- Durchflußsystems. Univ. Bremen, Fachber.1, Leistungsnachw.*

[3] *Föllinger, O., (1979). Regelungstechnik. AEG - Telefunken, 3. Auflage*

[4] *Popovic, D., G. Schulz-Ekloff, G. Thiele, E. Wendland, (1990). On Modelling, Identification, and Control of a Reactor for Pyrolysis of Organic Materials. Submitted to IFAC World-Congress, Tallinn, 1990*

Computer Coupled Laboratory Methanation Reactor for the Study of Temperature Profile Dynamics

D. Danziger

Abstract - The attenuation of temperature peaks in fixed-bed reactors is an important problem for safe process control of exothermic chemical reactions. As an alternative to conventionell methods of temperature peak attenuation the catalyst dilution has been studied. For the study of this method a computer coupled methanation reactor was developed.

1. Introduction

Reversible chemical conversions having high enthalpy changes might be used for a transfer of energy over long distances by recycling the reaction partners through pipelines. This, in principle, energy saving concept is studied in the "ADAM-EVA-Project" which uses the endothermic methane to synthesis gas conversion at the place of heat generation and the reverse exothermic methanation reaction for the energy consumers [1-3]. The broad application of such a system of energy supply requires safe, long-life methanation reactors needing minimum maintainance. The attenuation of temperature peaks for safe process control is generally realized by heat-exchangers, cold-gas injection, product feedback or multistage reactors [4-7]. Catalyst dilution might be another possibility for the reduction of temperature peaks [8-10]. For the study of the potential os such a measure a computer coupled methanation reactor was developed [11] and is described in the following, which permits adjusting catalyst dilution profiles and following changes of the axial temperature distribution.

2. Computer coupled methanation reactor

A schematic flow sheet of the computer coupled methanation system is shown in Figure 1. The mixed gases carbon monoxide and hydrogen (Messer-Griesheim) of H_2/CO ratio 3 taken from a gas cylinder, led through a flow-meter (1) to the reactor (4), preheated by an electric furnace (5), and converted in the catalyst bed (7). The product mixture containing CO, H_2, CH_4

and H_2O is led through a pre-cooler (12), a cooling trap for the removal of water (14) and a buffer vessel (16) to the outlet. The desired process pressure is adjusted at the reduction valve of the gas cylinder, and the feed rate is regulated by valve (13).

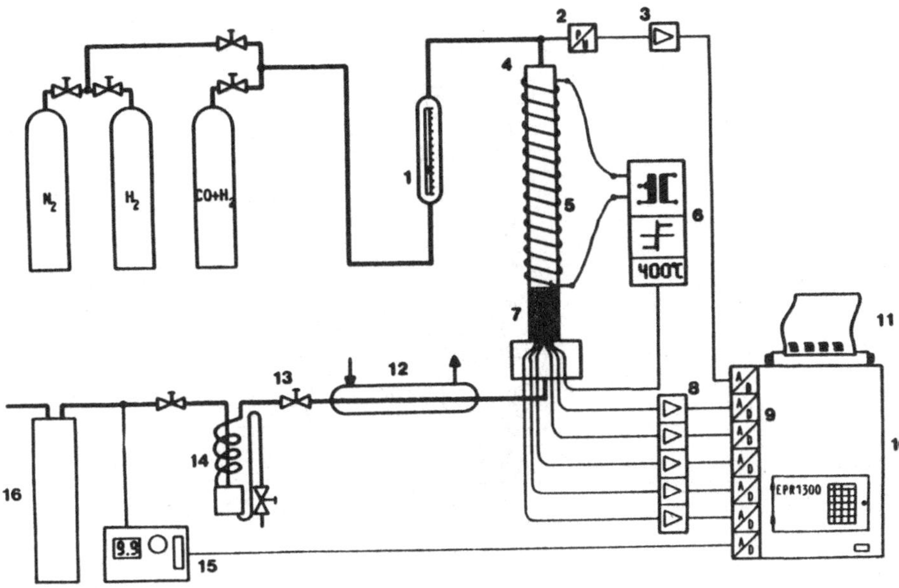

Fig. 1: Schematic flow sheet of the computer coupled methanation system. 1 flow meter (Rota), 2 pressure gauge (pressure-voltage-converter), 3 amplifier, 4 reactor, 5 pre-heater, 6 preheating controller, 7 catalyst bed with thermo-couples (Ni/CrNi), 8 amplifier, 9 analog-digital-converter, 10 process computer EPR 1300, 11 protokoll unit, 12 precooling (water-cooled), 13 valve for feed rate regulation, 14 cooling trap for reaction water, 15 CO measuring instrument, 16 buffer vessel

The laboratory methanation reactor (Fig. 2) consists of a V2A tube having a length of 50 cm and a diameter of 2 cm. The upper 40 cm of the reactor serve for the controlled preheating of the feed gas and have a winding of a heating filament (Philips). A thermocouple (Chromel-alumel) at the end of the preheating zone is used for the measurement and control of the feed gas temperature. The catalyst bed is placed in the lower part of the reactor and consists of a small basket with a length of 10 cm. Figure 3 shows a section through this part, where six thermocouples (Philips) with a length of 25 cm and a diameter of 1 mm are located and used for the measurement of the axial temperature profile. The reactor is wrapped up with ceramic fleece, glass-wool and dam material for its insulation.

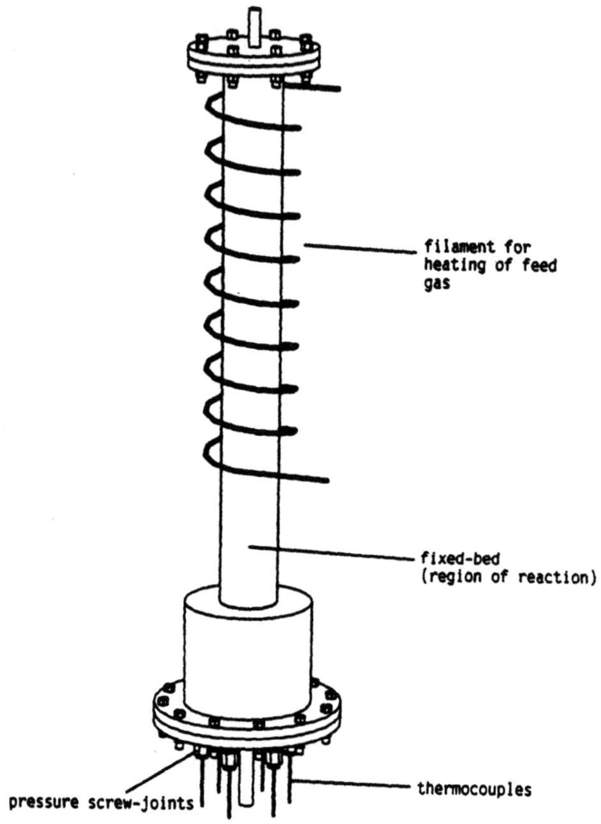

Fig. 2: Schematic presentation of the methanation reactor

The pressure, the temperatures in the fixed-bed and the CO-concentration at the outlet of the system are measured as process variables and are transmitted to a process computer (10). The CO-concentration is measured with the help of an infrared absorption instrument (Bosch). The process computer (EPR 1300, Krupp Atlas Elektronik) was programmed in the programming language PEARL under the multi-user operation system MOS.

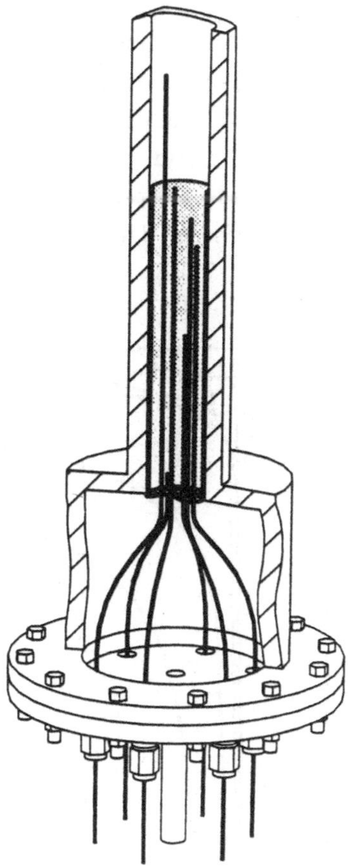

Fig. 3: Cut through the lower reactor part

3. Temperature profile dynamics

The undiluted fixed-bed consisted of 36 g of a pelletized (1 - 2 mm diameter) nickel catalyst (G-56H on Al_2O_3-support, Süd-Chemie AG). The following conditions were used: H_2/CO ratio=3, flow rate = 230 l h^{-1}, pressure = 0.26 MPa, feed temperature = 420 °C.

After 20 minutes of preheating under hydrogen the reactor was fed with the synthesis gas and the resulting dynamics are given in Figure 4. The ignition behavior, which can be followed clearly, causes a temperature jump about 400 °C. The ignition moves through the catalyst bed and measure point 5 has ignited after about 40 minutes. The experiment was finished by stopping the feed and all thermocouples indicated an exponential cooling behavior. A three-dimensional presentation of the temperature profiles, i.e. dependent on time t and on reactor length z is given in Figure 5. The axial temperature profile after 10, 20 and 30 reaction minutes is shown in Figure 6.

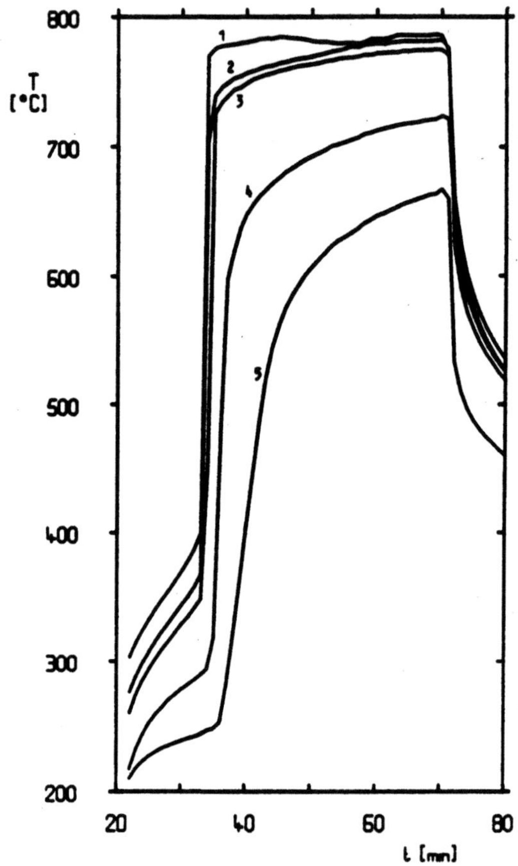

Fig. 4: Progress of temperature of measure points 1 to 5 dependent on time in the first experiment with fresh catalyst

The experiment was repeated twice with the same catalyst under the same reaction conditions. The deactivation of the catalyst can be followed by the temperature decline which is especially significant at the entrance of the fixed-bed (Fig. 7). The pressure had to be increased to carry out a further experiment with the required flow rate. Subsequently the reactor was disconnected and the catalyst was removed. An extensive carbon deposition on the catalyst was perceived.

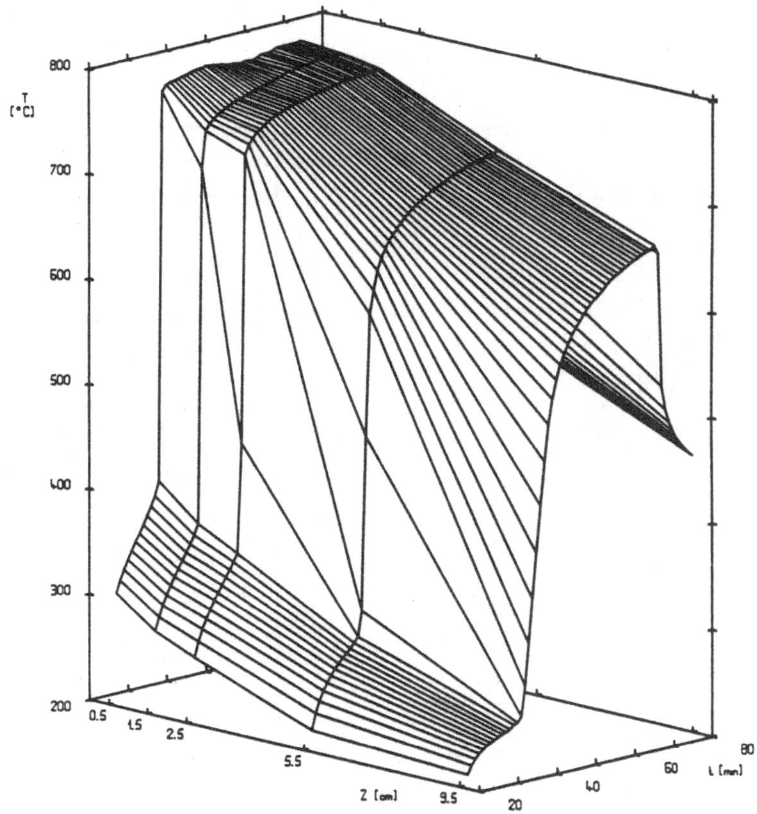

Fig. 5: Three-dimensional presentation of the temperature profile dependent on reactor length and on time in the first experiment with fresh catalyst

For the study of the effect of a catalyst dilution, series of experiments were carried out using different catalyst dilution profiles. Pelletized alumina oxide (Merck) was used as inert material and was mixed with catalyst pellets to the desired dilution which is characterized by the value η = catalyst / (catalyst + inert material) refering to volume units. A comparision of the measured temperature profiles with different catalyst dilutions is given in Figure 8. Increasing catalyst dilutions at the entrance of the fixed-bed gave decreasing temperatures. Furthermore more stable temperature profiles were obtained and an obstruction of the methanation reactor was prevented with catalyst dilution.

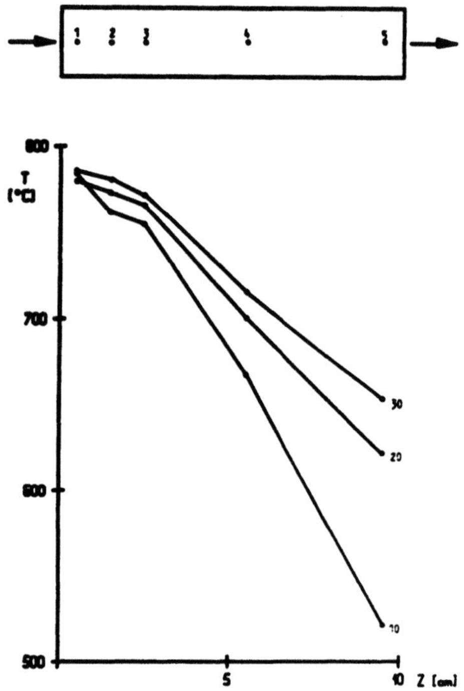

Fig. 6: Temperature profile of measure points 1 to 5 dependent on reactor length in the first experiment with fresh catalyst. The temperature profile is shown after 10, 20 and 30 reaction minutes

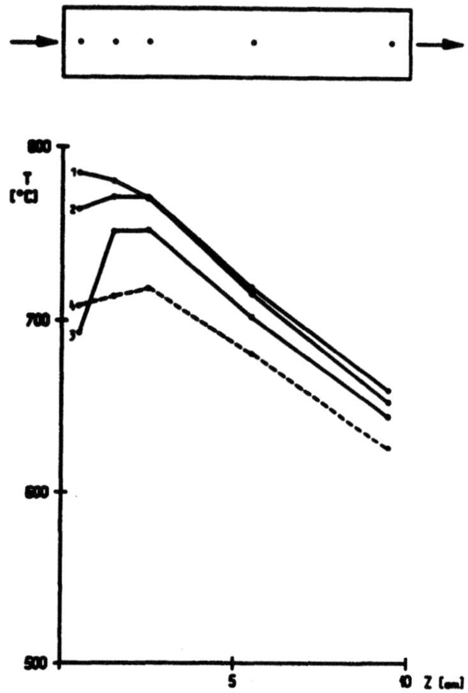

Fig. 7: Temperature profiles of four experiments with the same catalyst after 30 reaction minutes. The temperature profile of the fourth experiment was reached after a necessary pressure increase and therefore it is shown with dotted line.

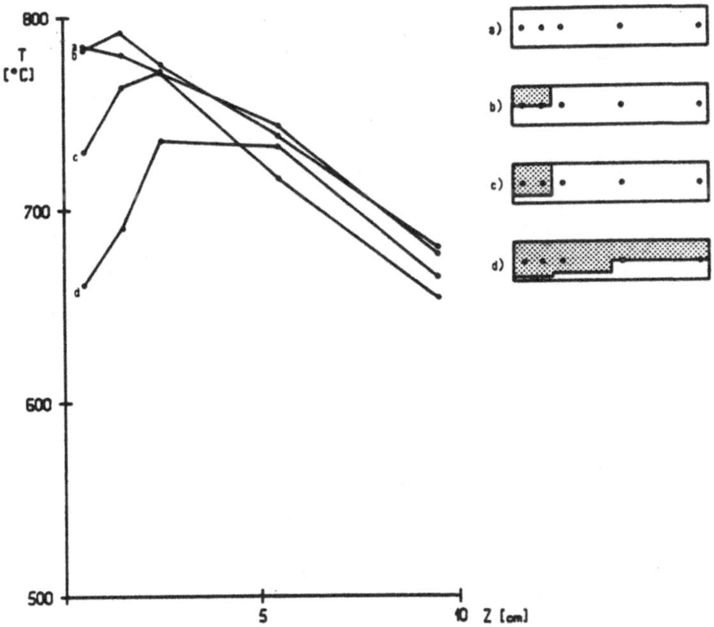

Fig. 8: Summary of experiments with and without catalyst dilution. The temperature profiles of the experiments with fresh catalyst are shown after 30 reaction minutes.

4. Thermodynamic of carbon deposition

For the evaluation of the most appropriate mean to reduce carbon deposition thermodynamic model calculations were carried out. The following simultaneous reactions were considered for calculations of equilibrium concentrations of carbon:

$$CO + 3\,H_2 = CH_4 + H_2O \qquad \Delta H_R^{\ominus} = -206.2 \text{ kJ mole}^{-1} \qquad (1)$$

$$CH_4 = 2\,H_2 + C \qquad \Delta H_R^{\ominus} = 74.8 \text{ kJ mole}^{-1} \qquad (2)$$

$$CO + H_2 = H_2O + C \qquad \Delta H_R^{\ominus} = -131.3 \text{ kJ mole}^{-1} \qquad (3)$$

$$2\,CO = CO_2 + C \qquad \Delta H_R^{\ominus} = -172.5 \text{ kJ mole}^{-1} \qquad (4)$$

The computations of the simultaneous reactions were realized with a computer program according to Henley and Rosen [12] which was implemented on the process computer . Fig. 9 represents carbon fractions dependent on the temperature with a gas composition of 25% CO and 75% H_2 and a pressure of 0.26 MPa if reactions (1) to (4) were taken into account. The carbon concentration has a thermodynamic maximum value at 600°C.

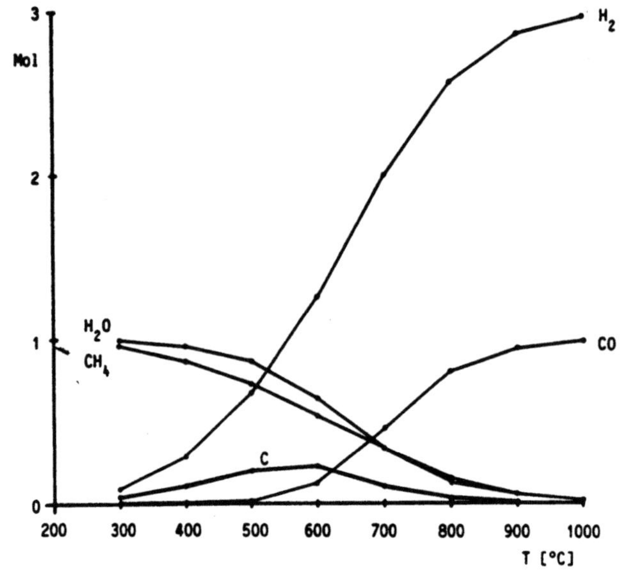

Fig. 9: Equilibrium concentrations of methanation gas ($H_2/CO=3$ P=0.26 MPa) dependent on the temperature if the following simultaneous reactions were taken into account:
$CO+3H_2 = CH_4+H_2O$, $CH_4 = 2H_2+C$, $CO+H_2 = H_2O+C$, $2CO = CO_2 +C$

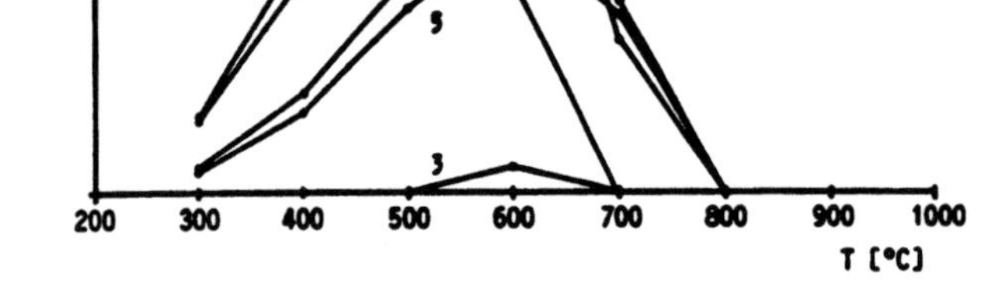

Fig. 10: Comparision of equilibrium carbon fractions
(1) P = 0.26 MPa, 25% CO, 75% H_2
(2) P = 0.26 MPa, 22.2% CO, 66.7% H_2, 11.1% H_2O
(3) P = 0.26 MPa, 20% CO, 80% H_2
(4) P = 2 Mpa, 25% CO, 75% H_2
(5) P = 3 MPa, 25% CO, 75% H_2

A reduction of carbon formation can be obtained by (i) steam-injection, (ii) increase of pressure and (iii) increase of the H_2/CO ratio. A comparision of these methods is given in Fig.10 and shows that the increase of the H_2/CO ratio is the most effective one.

5. Outlook

A computer coupled methanation reactor was developed which permits adjusting variable profiles of catalyst dilution and measuring dynamics of the axial temperature profile. This dynamic depends on the ignition behavior as well as on depositions of carbon. The thermodynamics of the carbon formation suggests continuing the investigations with gas compositions $H_2/CO > 3$. The concept of the catalyst dilution can be analyzed in connection with other methods for the attenuation of temperature peaks. If stationary conditions can be achieved, a reaction model of the methanation might be developed with the help of identification methods, which can be used for computer simulation and process optimization.

References

[1] H.Harms, B.Höhlein and A.Skov
Chemie-Ingenier-Technik, 52, 1980, p.504-515

[2] H.Harms, B.Höhlein, E.Jorn and A.Skov
Oil and Gas Journal, 78, 1980, p.120-127

[3] J.Range and H.G.Harms
Chemie-Technik, 11, 1982, p.894-899

[4] G.Eigenberger
Chemie-Ingenieur-Technik, 51, 1979, p.1105-1110

[5] G.K.Giger, R.Mutharasan and D.R.Coughanowr
Ind. Eng. Chem. Fundam., 19, 1980, p.389-396

[6] D.C.Hempel
Chemie-Ingenieur-Technik, 48, 1976, p.71

[7] L.D.Gaines
Chemical Engineering Science, 34, 1979, p.37-50

[8] P.H.Calderbank, A.Caldwell and G.Ross
Proc. 4th Europ. Symp. Chem. Reaction Eng., Brüssel 1968, Pergamon Press, Oxford 1971, p.93-106

[9] D.Danziger, D.Popovic and G.Schulz-Ekloff
Canad. J. Chem. Eng., 61, 1983, p.126-128

[10] D.Danziger and G.Schulz-Ekloff
Informatik-Fachberichte , Springer Berlin 1983, p.562-567

[11] D.Danziger
Disssertation, Univ. Bremen 1984, Fortschr.-Berichte VDI, Reihe 3, Nr.102, VDI-Verlag, Düsseldorf 1985

[12] E.J.Henley and E.M.Rosen
Wiley, New York 1969

5
Expert Systems

Some Aspects of an Architecture for Real-Time Expert Systems in Industrial Plant Automation and Supervision

Walter Fedderwitz

Abstract

The highest level of control and supervision of industrial plants, for instance fossile or nuclear power generating systems or chemical production units, still has to be done by human operators. To support these operators, there is an increasing demand for real time and on-line expert systems.

Most of the expert systems in industrial use today are of the interactive type. They work on static situations and query the user to gather information. But when shifting to a time-varying environment, where the expert system is connected to the plant and incoming data change while the expert system infers, new architectural concepts of these systems are needed.

By watching and analysing the behaviour of human operators in different plant stati, we developed an architecture for real-time expert systems centered around the situation concept. Not incoming data, but abstractions of these data (in it's own a knowledge based operation), called situations, are the basis for human decisions in plant operation. This concept leads to a multi-parallel expert system architecture to be described in the paper.

1.) Industrial Plant Control as Application Field for Real Time Expert Systems

Monitoring and control of industrial plants is a task which is found in nearly all industrial production facilities. On the one hand it covers functions such as controlling individual subsystems and on the other hand we find the control of the complete plant at a high level of abstraction. The primary task is to coordinate information about state changes in the plant, messages sent by automatic subsystems in the plant, and measurement values.

The common objective of process control is /1/

- to increase safety of operation and
- to increase economy.

By introducing expert systems into these tasks, we do not expect to replace the proven automation system but to supplement them, because there are a lot of problems and tasks which are hard to solve in a conventional approach. In particular, load management and fault management contribute to these problems /2,3,4/. Today these problems are solved by operators with varying success related to their experience.

2.) Common Requirements for an Expert System in Industrial Control

The idea of having an expert system supporting a plant operator identifies a lot of problems:

o The system has to do time reasoning, because signals from the plant vary over time.

o Answers have to be available within certain time intervals or they become useless.

o The system has to give advice not only in particular events, but offer continuous support for diagnosis, initial treatment of fault, repair, return to normal operation.

o The system must not lock up by one problem, but needs the power to deal with different problems at the same time. Relationships between these problems possibly overlooked by the operator should show up.

o The system cannot rely on data entering by the operator, because preventive actions mean to monitor data continuously.

3.) The Architecture

The classical expert system of the interactive type acts on static situations, not varying over time. Typically the operator sits down at the system and enters the data relevant to the problem at hand by answering questions. The system offers an inference result and at the end of the session the system is brought back into its initial status and is ready for the next job.

This interactive type of an expert system already can do time dependant reasoning. As an example, every diagnosis looking not only at the current state of the plant, but also mentioning the plant status in the past, does time dependant reasoning in the sense that it argues over time, time intervals and causal relationships.

The difference between real-time expert systems and this type of dialog oriented systems is twofold:

- The real-time system has to deal with newly arriving information while it is inferring because the plant normally cannot stop operation.

- Inference results have to be available within certain time constrains given by the dynamics of the technical process. Results which are correct but only late in time are worthless because the plant status may have changed in such a way that for example corrective actions as part of the inference result are coming too late.

This leads to the following definition:

Real Time Expert System

A Real Time Expert System is an expert system producing answers within a pre-known time interval related to the dynamics of the plant without exhausting a finite amount of resources (memory, cpu-power).

3.1) The Digital-Controller-Type Architecture

One approach to develop an architecture for real-time expert systems extends the concept of digital controllers into expert systems.

Digital controllers act on sampled versions of the incoming sensor data. Under the assumption that the analog signal is band-limited, the complete information is already in the samples. For this mathematical concept see e.g. /5/.

The code of the controller is written in such a way, that it runs once whenever a new sample arrives, see figure 1. This approach allows to consider the complete information in the incoming signal.

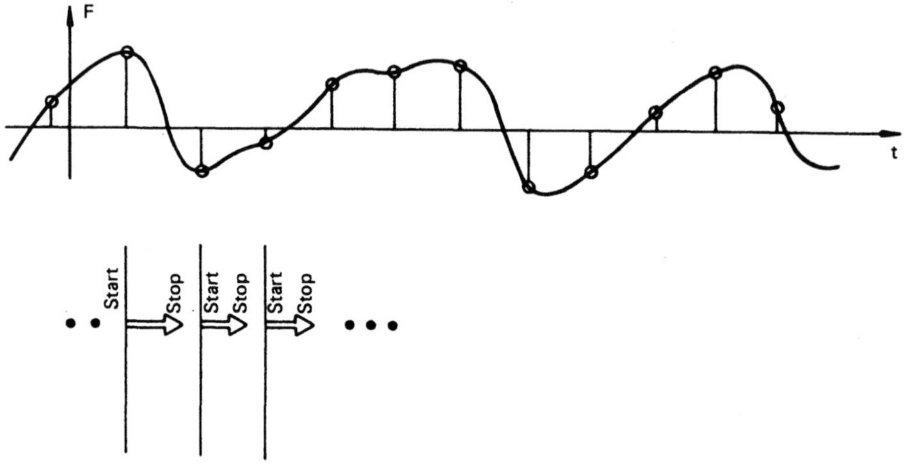

Fig. 1 The Digital Controller

While this approach has been proven very successfull in the area of controllers, it is difficult to extend it into expert systems. When we follow this line of thought, it means to design an expert system which runs completely through its knowledgebase following every possible path once for every incoming sample of data. Often data samples arrive at rates of tens to thousands per second or ever higher. This leads to two problems. Computers to allow this approach have to be very fast and therefore expensive.

The second problem is a principal one. Humans, that is to say, plant operating personell, do not infer and analyse like digital controllers. This fact is proven easily in the phase of knowledge acquisition. Interviews show that experts are following different approaches to be described in the next paragraph. If one nevertheless decides to follow the controller-type architecture for an expert system, it is not possible to acquire information from the experts. The knowledge in this type of systems has to evolve elsewhere.

3.2) The Human Model Paradigm

An alternative way to develop an architecture for real-time expert systems emerges from close observation and analysis of human (e.g. plant operator) behaviour. The human operator faces exactly the same two problems as the real-time expert systems: New data arrives while inference processes are running and answers have to be available in certain time intervals.
The operator behaviour falls into two different main categories, observation and deviation treatment.

Observation

Most of the time the plant runs along the predetermined way, without any faults. The operator continuously monitors (part of) the incoming data to find hints for beginning disorders. The observation cycle looks like the controller-type approach because it may happen that the observation cycle has to run once for every incoming sample of data. But there is a main difference. The decisions taken in the observation loop are relatively simple compared to the complete analysis of a disorder, because they only lead to the initial ideas of things going wrong. Typical analysis in this stage looks like: "That value is rising fast!" or " That alarm went off!".

Fault treatment

Analysing the human behaviour in this case is the key element to the system architecture searched for. Shortly after detecting a deviation from expected plant operation, the operator in his or her mind builds up an initial idea or guess of what might have happened. We call this guess the initial <u>situation</u>. Now operator actions fall into two different time lines as shown in figure 2.

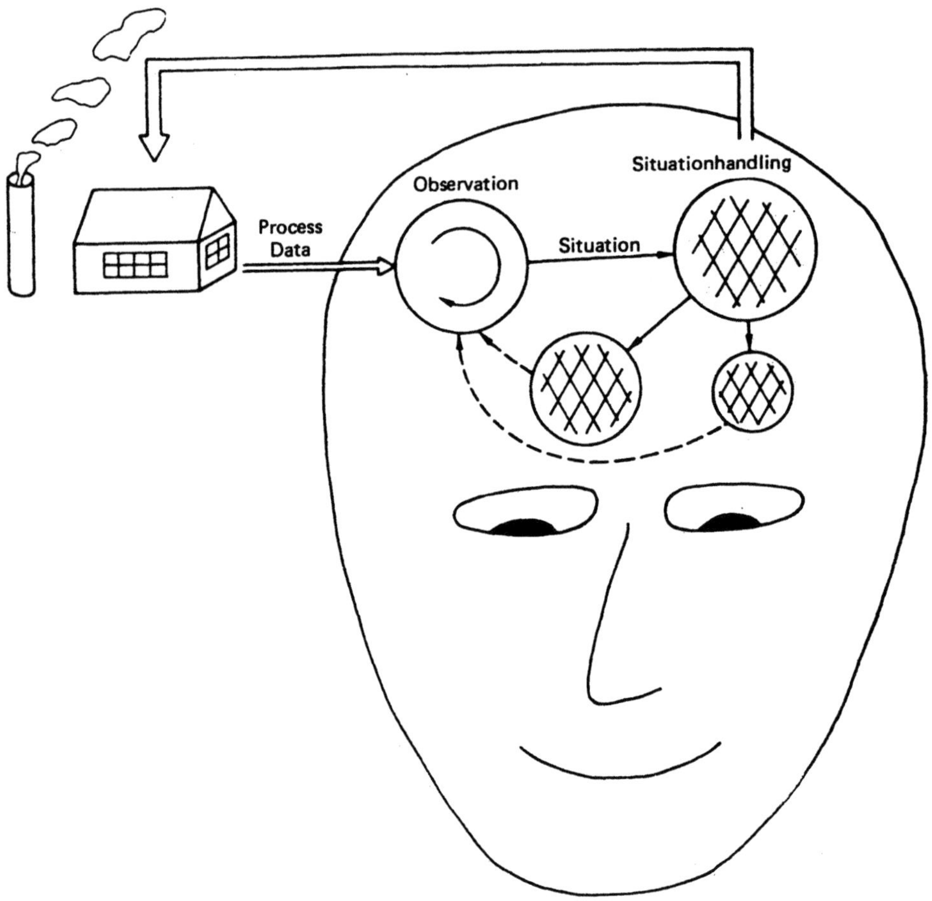

Fig. 2 Situations In The Human Mind

o Operator Activity According to the Situation

When the assumed situation comes close to the actual things happening in the plant, the situation will be valid for a long time compared to the sampling rate of the incoming data. The operator now decides **on the basis of the situation** on the next steps of his or her intervention. Because the situation is stable for some time, the decision to be taken needs not to be done in the close time contraints of the incoming data. The time constraints are somewhat relaxed and the information (the situation) to decide on is fixed. We call this a relative decoupling of the inference process from real-time constraints. The decoupling is relative in the sense that we only gain some time, not an arbitrary amount. The inference based on the situation has to finish before the situation becomes invalid. Figure 3 shows the time relationships. A good estimate for the time available is the time the expert uses in this same situation.

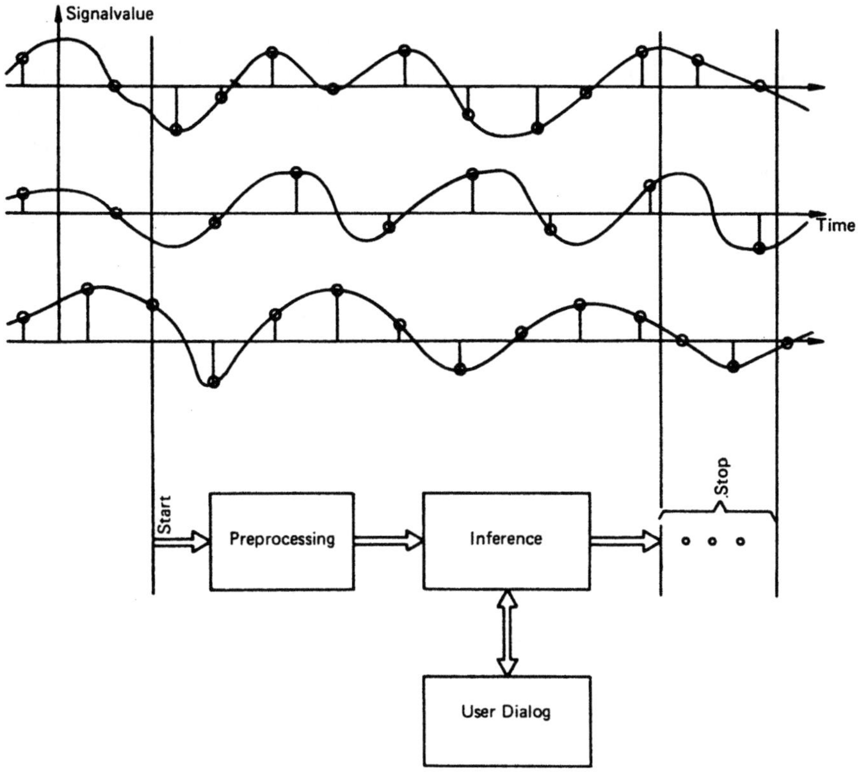

Fig. 3 Start And Duration Of A Situation

As the plant status is still changing, invalidity of the situation becomes a crucial point in this concept. Perhaps we can tell in advance the period of time the situation is valid. But we cannot be sure, because something unexpected may happen. This raises the need for the second line of operator activity, he or she is solving in parallel with the inferencing described above.

o Detecting Unexpected Invalidity of a Situation

Parallel to the main activity of acting upon the situation, the operator keeps an eye on the incoming signals. He or she now decides whether something unexpected happens invalidating the situation or whether the signals still fit into the situation.

As long as the situation is valid under these observations, the inferencing based on it proceeds. When the situation (unexpectedly) becomes invalid, the operator has to start again by building up a new situation. This idea of <u>situations</u> and the various parallel "lines of thought" in time are illustrated in figure 4.

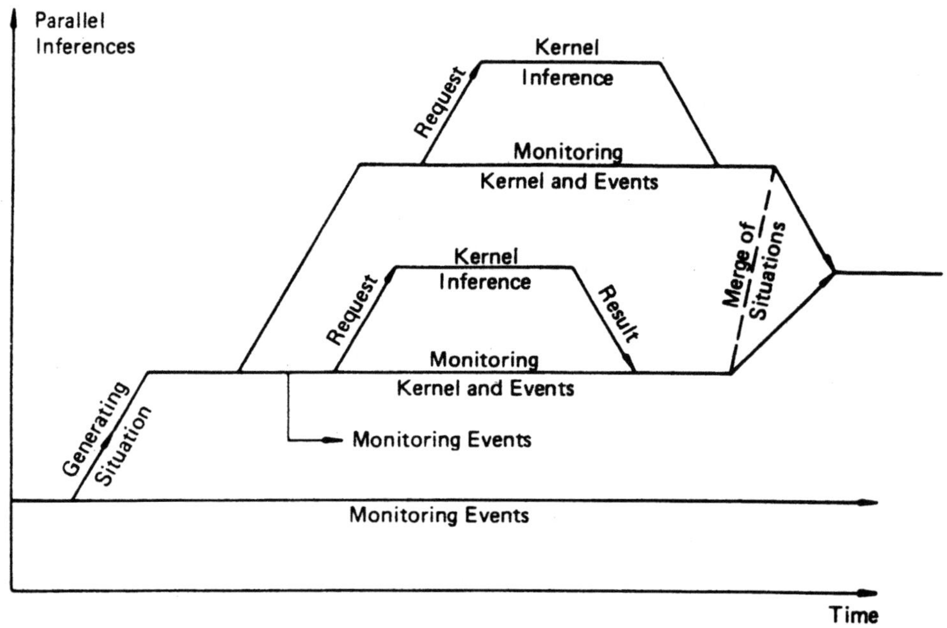

Fig. 4 Parallel Inferences In Time

We see that a situation does not necessarily use all available incoming information. When the knowledge about the beginning of a situation and the following reasoning is correct, already a part of the information will do. This fact is proven by every human expert operating successfull under exactly these circumstances.

When situations became invalid unexpectedly, the system as well as the human operator may run into severe trouble because there is no chance to close up to the process dynamics. A good design will avoid this as far as possible. But there remains a reisk. It is our strong opinion that expert systems can be used in plant automation only when this rare events are catched by other layers of the automation system. To clarify this look at a nuclear power plant, which normally follows these design goals to catch operator errors. Normally the operator will act according to the situation and a beginning malfunction turns back into normal operation. But when there is no or wrong operator action, the low level control system will shut down the plant and thus fail safe. These concepts are further discussed in /6/.

3.3) The System Layout

The real-time expert system resembles the structure of operator intervention described above. The architecture falls into three main building blocks, the signal world, the situation world and the kernel world (figure 5).

3.3.1) Signal World

This first building block is the one closest to the incoming data and thus to the conventional plant automation. As we stated in the beginning, we do not attempt to replace the conventional computer system, but see the expert system as an add-on part covering tasks solved before by humans only. The task of the signal world is to receive incoming data and have this data available for further activities. The signal world is at the border between the conventional automation system and the other parts of the expert system. An object oriented approach is very successfull at this stage. Every signal is modeled as an instance of an object SIGNAL and typically has slots to keep the actual value of the signal, the receiving time and the history of these values. Furthermore there is a link into the plant description allowing the expert system to reason about this signal in conjunction with the plant description and other signals.

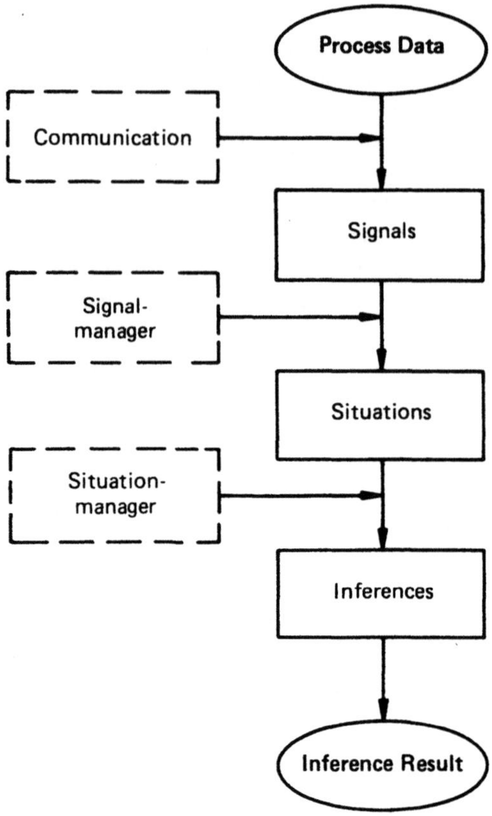

Fig. 5 General Outline Of A Real Time Expert System

Whenever a new sample has arrived, the signal-object informs all receivers which are interested in this new information. The receivers are situations, to be described in the next section.

3.3.3) The Situation World

The situation world is the next and central building block of the system. Syntactically this world is similar to the signal world, the signal objects being replaced by situation objects. The difference between the two worlds is in the semantics. The situation objects have to model the situations the operator builds up in mind. As a starting point we have one primary situation modelling the operator in his observation cycle.

Whenever the observation detects deviations from expected plant behaviour, a new situation becomes active describing this problem. Both situations run in parallel allowing to go on with observation while analysing. In case of unexpected changes in the plant, the situation may be dropped and the inference has to start again.
When another problem or subproblem is detected, other situations are active in parallel allowing the system to deal with these problems.

The initial situation is an example of how the system can be better then the operator without sacrifycing the idea of modelling human behaviour. The operator may tell us that in principle it would be helpful not only to monitor the signals he actually monitors, but also some other signals. The operator often cannot do this at every time because of his limited capabilities. The system can easily be set up to do this continuous monitoring.

3.3.4) The Kernel World

The kernel world allows a situation to infer in parallel mainly to the observation cycle, but also to other situations. The split into kernel world and situation world is primarily driven by implementation aspects, not so much because the human behaviour leads to this split. Every kernel allows for one "chain of thoughts" related to a situation. A kernel is implemented as a computing process of its own and therefore runs "parallel" to other processes on the machine.

This expert system kernel is a complete expert system as we know it from dialogue-oriented systems. The kernel starts with the initial information supplied by the starting situation.

The kernel is allowed to query whatever object it likes to ask: other situations and signals, the database describing the plant or possibly the user. The kernel builds his own local database keeping his results. This local data is not available outside the kernel. When the kernel finishes, it hands back the result to the associated situation. Only then a modification of global data structures may take place. The inference result for instance may create and start an instance of some special situation frame dealing with a certain type of fault.

By restricting the kernel to change global objects only via results returned to the situation, we give the situation a chance to decide whether to use the result or not. The situation will not use the result when, in parallel to the run time of the kernel new information arrived which changed the state of the situation. This is the unexpected case where the operator by keeping an eye on the data arriving discovers that his ideas were wrong. A more extended discussion of these problems can be found in /6/.

4.) Future Steps

The first real-time expert systems following the idea of the situation concept now undergo field trials, in which it is connected to a real technical process. The architecture of real-time expert systems will for shure become more elaborate from this experience, but first results show that our concept is usefull.

Literature

/1/ Zentrale Leittechnik, in: Schriftenreihe "Prozeßrechner in der Praxis", Band 9, Krupp Atlas Elektronik, Bremen, 1980

/2/ Eiben, B. Eisermann, J. Fedderwitz, W.: PARES Ein Expertensystem fuer die Leitwartentechnik
Conference of the German Chapter of the ACM, Nuremberg, 1987

/3/ Wittig, Th. The KRITIC Project: ESPRIT supported Project on Industrial Control
In: Mamdani,Efstathiou (Eds) : Expert Systems and Optimization in Process Control, Technical Press, 1985

/4/ Wittig, Th. The KRITIC Project : Domain 2 : Controlling complex Power Distribution Networks
BIRA/ACM Seminar on Artificial Intelligence in Industrial Applications, Antwerpen, 1986

/5/ L.R. Rabiner, B. Gold, Theory and Application of Digital Signal Processing, Prentice Hall, New Jersey, 1975

/6/ Kerndlmeier u.a. (Hrgs.) Technische Expertensysteme fuer Diagnose und Prozessfuehrung, Oldenbourg-Verlag 1989

Rule-Based Modelling of Dynamical Systems

H. Kiendl, M. Krabs and M. Fritsch

A new rule-based method for the modelling of dynamical systems is summarized and further developed. It allows to model the input-output system behaviour by IF-THEN-rules. The modelling is based on recorded time functions of input and output variables and generalizes the observed behaviour, even if these observations are incomplete, inconsistent or noisy. Resulting inconsistencies among the obtained rules are resolved by an inference strategy. The method can be used for the modelling and design of decision strategies in discrete event systems and of controllers for continuous systems. As an application example the design of a rule-based two-level controller is described.

1 Introduction and survey

The modelling of dynamical processes is basic to their analysis, understanding, simulation and manipulation. Difference and differential equations are appropriate for modelling great classes of technical systems and such models can be treated analytically and numerically in an efficient way by a well developed theory.

Beside this "classical" approach the modelling of dynamical processes by IF-THEN-rules has found increasing interest in the recent years. On the one hand efficient computer assisted tools for analysis, simulation and verification of such models have been developed. On the other hand many technical systems – even if complex, discontinuous or unknown in nature – show an input-output-behaviour, which can be modeled by rules. The challenging problem of finding an appropriate set of rules is usually solved either by interviews with human experts [1] or by learning from observations. The second method is conventionally based on learning disjunctive normal forms, interpreting each single observation as a rule. Some experimental approaches in KI-research provide more flexibility with respect to the type of the rules [2,3]. Incomplete, inconsistent or noisy observations however are still an unsolved problem, especially in the modelling of dynamical processes.

In this paper we summarize and generalize a new interactive method for the generation of rule-based dynamical models, based on observations

[4,5,6]. Furthermore we design a rule-based controller as a new application example.

The method generalizes from examples even if these are incomplete, inconsistent and noisy. Thus the generated models are not necessarily completely correct, which can be often tolerated. For the application of the method an inductive expert system has been developed which supports the user. It allows to structure the admitted rules by making use of heuristic or physical preknowledge.

The efficiency of the method, so far established for relatively simple dynamical systems, motivates further applications on more complex systems. Especially discrete event systems seem to be a promising application field. In high level process control for example, condensed and discretisized input information is available and the control output consists of decisions on discrete operations. In resource scheduling and operation planning suitable actions are often based on observable discrete indicators. Control rules then can be learned from operation data records or from simulation.

2 Modelling concept

2.1 The discrete event substitute system

We consider dynamical systems with input variables $x_n(t)$ and output variables $y_m(t)$, the time functions of which are empirically known in a given time interval $0 \leq t \leq T_0$. To model the characteristics of interest a discrete event substitute system with *input events* e_i and *output events* a_j is introduced using heuristic or physical preknowledge which is available about the system behaviour and the modelling context (fig.1). The substitute system is based on the specification of

- qualitative characteristics of the input variables $x_n(t)$ and of the output variables $y_m(t)$ which are potentially relevant in the modelling context and therefore shall be regarded as input events e_i and output events a_j. Each event may depend on several input variables $x_n(t)$ respectively output variables $y_m(t)$ and vice versa from each variable $x_n(t)$ or $y_m(t)$ more than one event e_i respectively a_j can be derived.

- corresponding quantitative *decision functions* which can be applied to the observed time functions $x_n(t)$ respectively $y_m(t)$ and map them to the three possible values "1", "0" and "?" of e_i and a_j meaning

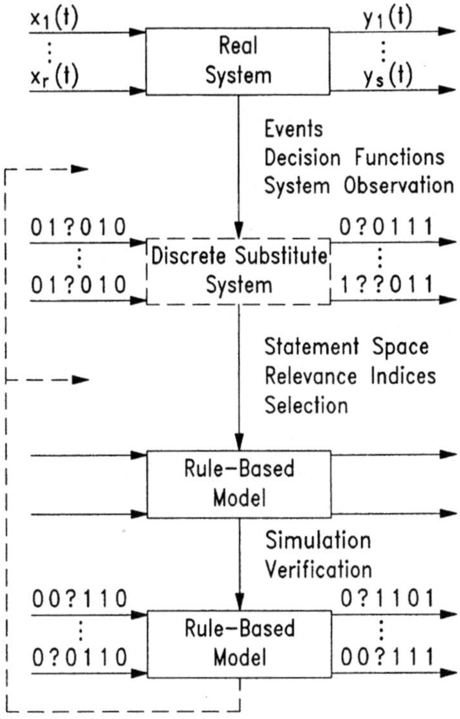

Figure 1: Here proposed concept for generation of rule-based models

$$\begin{array}{lcl} 1 & := & \text{"event is } \textit{PRESENT}\text{"} \\ 0 & := & \text{"event is } \textit{NOT PRESENT}\text{"} \\ ? & := & \text{"presence is } \textit{UNDECIDABLE}\text{"} . \end{array} \quad (1)$$

- an appropriate sequence of instances k, at which the observed time functions $x_n(t)$ and $y_m(t)$ have to be investigated by application of the decision functions.

2.2 System observation

The required records of $x_n(t)$ and $y_m(t)$ can be obtained by observing the interesting real or simulated system during normal operation or by observing experiments which cover a greater amount of relevant system states. The application of the decision functions on these records of $x_n(t)$ and $y_m(t)$ yields the sequences $e_i(k)$ and $a_j(k)$. These are the *input* and *output sequences* of the discrete event substitute system consisting of the elements "1", "0" and "?" (fig.1).

2.3 Specification of the statement space

In "classical" modelling usually the model structure with free parameters has to be specified by the use of physical or heuristic preknowledge. Analogously here we define a space of *statements* which determines structure and complexity of the rules that seem to be adequate in the modelling context. A statement $A_{\lambda j}$ of this space describes coincidences of an *input situation* E_λ and an output event a_j and has one of the following forms:

$$\begin{array}{l} A_{\lambda j}[1] \text{ is defined as} : E_\lambda = \text{TRUE and } a_j = 1 \\ A_{\lambda j}[0] \text{ is defined as} : E_\lambda = \text{TRUE and } a_j = 0\,. \end{array} \quad (2)$$

At each instant k it can be investigated which value a_j has, which input situations E_λ are TRUE and thus which of the statements hold. An input situation E_λ is defined by specification of a *window* and a *window condition*. A window has a well-defined shape and a well-defined delay D with respect to the reference instant k and thus selects certain elements of the input sequences (fig.2). Window conditions refer to the occurance of the

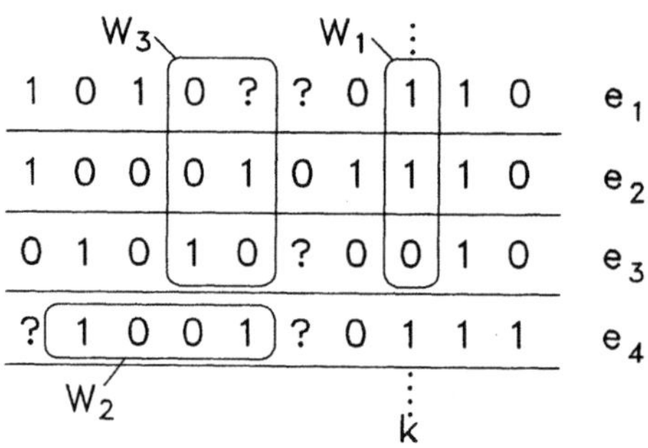

Figure 2: Examples for windows which are used for the specification of input situations. The windows W_1, W_2 and W_3 have the delays $D = 0$, $D = 3$ and $D = 3$ with respect to the reference instant k

elements "1", "0" and "?" within the window. Examples for such conditions which can be relevant in the modelling context may for example be:

- There is a specific pattern of the elements "1" and "0"
- There are more than p elements "1" (or "0")

within the window. If the elements in the window meet the window condition at an instant k then we say "$E_\lambda(k) =$ TRUE".

The complexity of the resulting *statement space* depends on the considered windows and window conditions. For the application it is important to define the statement space in such a way that on the one hand enough modelling flexibility is provided and on the other hand combinatorical explosion is prevented.

2.4 Search in the statement space

The modelling of the substitute system – given by the input and output sequences $e_i(k)$ and $a_j(k)$ – is based on the automatical investigation of all statements of the statement space (complete search) or – to save computing time – only of the most promising statements, selected by a heuristic strategy (incomplete search). For each investigated statement $A_{\lambda j}$ the numbers

$$\begin{aligned} N(A_{\lambda j}[1]) &:= \text{number of instances } k \text{ at which } A_{\lambda j}[1] \text{ holds} \\ N(A_{\lambda j}[0]) &:= \text{number of instances } k \text{ at which } A_{\lambda j}[0) \text{ holds} \end{aligned} \tag{3}$$

are determined. From these numbers we derive the following three indices, which are associated with each considered statement.

- The *situation number*

$$S_{\lambda j} = N(A_{\lambda j}[1]) + N(A_{\lambda j}[0]) \tag{4}$$

  indicates how often the input situation E_λ has been observed while the output event a_j was either "PRESENT" or "NOT PRESENT".

- The *quota of events*

$$Q_{\lambda j} = \frac{N(A_{\lambda j}[1])}{S_{\lambda j}} \cdot 100\% \tag{5}$$

  is the conditional frequency with which the output event a_j is observed when the input situation E_λ is TRUE. It is interpreted as a measure for the safety (probability) with which the output event a_j can be predicted if E_λ is TRUE at the instant k.

- The *relative output event frequency* H_j is defined as percentage of the number of instants k for which $a_j = 1$ holds in relation to all instances k for which $a_j = 1$ or $a_j = 0$ holds. This percentage is a measure for the probability with which the output event a_j can be predicted if we don't consider the inputs.

2.5 Valuation of statements

The above indices are used in the following heuristically motivated and empirically tested way to estimate, to which extend it is justified to interpret a statement as a rule:

- Clearly positive values of $Q_{\lambda j} - H_j$ point to a correlation between the input situation E_λ and the presence of the output event a_j. This is a hint for the rule

$$\text{IF } E_\lambda(k) = \text{TRUE THEN } (a_j(k) = 1). \tag{6}$$

- Clearly negative values of $Q_{\lambda j} - H_j$ point to a correlation between the input situation E_λ and the absence of the output event a_j. This is a hint for the rule

$$\text{IF } E_\lambda(k) = \text{TRUE THEN } (a_j(k) = 0). \tag{7}$$

- Small values of $|Q_{\lambda j} - H_j|$ indicate that there is no significant correlation between the input situation E_λ and the output event a_j.

In addition to this the extend of the empirical base indicated by $S_{\lambda j}$ has to be taken into account. Statements which are confirmed often by the given sequences should be valued higher than statements with input situations that occured seldom because latter represent rather random than systematic behaviour.

Therefore the following two different *relevance indices* $J_{\lambda j}$ and $J^*_{\lambda j}$ are introduced to value the relevance of a statement. The index

$$J_{\lambda j} = (Q_{\lambda j} - H_j)\sqrt{S_{\lambda j}} \tag{8}$$

stresses the extend of the empirical base of a rule. The index

$$J^*_{\lambda j} = (Q_{\lambda j} - H_j)\sqrt[4]{S_{\lambda j}}. \tag{9}$$

stresses more the aspect of safety of a rule.

2.6 Selection of a rule set

To generate an appropriate set of rules the user has to specify minimum requirements which the above indices have to meet:

- Demanding that the quota of events $Q_{\lambda j}$ is between $u\%$ and 100% he can select rules of the type (6) which are the safer the larger u is chosen. Demanding that $Q_{\lambda j}$ is between 0% and $v\%$ he can select rules of the type (7) which are the safer the smaller v is chosen. For the choice of u and v the value of H_j should be taken into account.

- By choosing a lower bound for the situation number $S_{\lambda j}$ he can select rules which are based on a reasonable number of observations.

- Furthermore the statements $A_{\lambda j}$ are grouped according to the outputs a_j and the inputs e_i which are involved in the input situation E_λ. Then the best rules according to one of the relevance indices J and J^* are selected from each group.

Finally logically dependent rules can be eliminated: Each rule, which is a special case of a more general and better rule, is removed.

Usually the resulting rule set is not free of contradictions. This is a necessary consequence of the generalizing feature of the modelling concept and of the fact that incomplete or noisy system information is admitted for the rule generation.

2.7 Inference strategy

An inference strategy is used to obtain the output of the rule-based model. It resolves possible contradictions among all rules, the situation E_λ of which is TRUE at the actual instant k (applicable rules). A very simple but often successful inference strategy consists in taking the rule with the best relevance index (Best Expert Strategy). Another effective strategy takes several applicable rules into account, selected and weighted by the corresponding relevance indices (Expert Council Strategy). Moreover it seems to be promising to take into account also the degree of generality of a rule as well as logical connections between rules.

2.8 Verification of the generated model

It has to be checked to which extend the model shows the same response to given input sequences as the underlying discrete substitute system. This depends mainly on the completeness of the observed input events, on the adequacy of the chosen rule types and on the criteria chosen for the rule selection. If necessary the corresponding steps in the modelling process have to be modified (broken arrows in fig.1).

Furthermore it has to be examined to which extend the model describes inherent regularities of the modelled system instead of only random correlations between the given input and output sequences. The larger the chosen statement space is in relation to the length of the observed input and output sequences, the more probable it is according to statistical hypothesis test theory that observed correlations are only random in nature. Therefore it is necessary to compare the behaviour of the model with additional input and output sequences of the discrete substitute system which have not been used for modelling.

2.9 Automatic improvement and adaptivity

Starting with a rule set the associated indices J and J^* can be updated automatically using new sequences obtained from system operation. The most successful rules improve their indices and win in the inference process. Thus the model becomes more and more accurate. Based on the same technique also model adaptivity to a slowly time-variant real system can be achieved: The numbers (3) are multiplied by a "forgetting factor" q, the value of which is chosen between 0 and 1, adjusted to the expected speed of change.

3 The expert system ROSA

To apply and test the above modelling concept we developed the inductive expert system "ROSA" (Rule Oriented System Analysis) which supports the user to perform the steps described above [6]. In ROSA the following specifications of the modelling concept are realized:

- Windows enclose either several instants of one input sequence (single input windows) or several inputs at one instant (single instant windows) as for example the windows W_2 respectively W_1 in fig.2.

- Maximum values D_{max}, T_{max} and G_{max} can be specified for the *window parameters*:

  - delay D

- number T of instants enclosed
- number G of inputs involved.

- Window conditions have one of the forms
  - "The window shows the following pattern of the elements 1 and 0".
  - "The window shows exact / less than / more than p elements 1 (or 0)".
- Complete search is performed in the rule space.
- As inference strategies the "Best Expert Strategy" and the "Expert Council Strategy" are realized.

3.1 Illustrative Example

We assume that a fictitious preservative causes allergic illness according to the following regularities:

A. Reception of the preservative twice within three days causes allergic illness at the day of the second reception.

B. Reception of the preservative and sugar during the same day causes allergic illness immediately.

We assume that preservative and sugar are contained in five kinds of food according to table 1.

Contains \ Food	Cheese	Milk	Juice	Cake	Peaches
Preservative	yes	yes	no	yes	no
Sugar	no	no	yes	yes	no

Table 1: Assumed occurence of the two allergy causing substances preservative and sugar in five kinds of food

An accidental schedule of meals has been set up for 50 days and it has been noted whether allergic illness has broken out according to the regularities A and B. The resulting schedule corresponds to a substitute system with five input sequences and one output sequence. To analyse these sequences with the expert system ROSA we specified the statement space by admitting single input and single instant windows restricted by $T_{max} = 3$, $G_{max} = 2$ and $D_{max} = 0$. As relevance index we chose J^*.

With this specifications ROSA generated 181 statements and sorted them according to the index J^*. To get "safe" rules we selected all rules with $Q_{\lambda j} = 100\%$ and so obtained 26 statements. After removal of logically dependent statements 10 rules remained (table 2). They describe the regularities A and B correctly and completely.

RULE 1	IF an event occurs at the input cake – THEN an event occurs at the output allergy. S = 9; Q = 100.0 %, J^* = 97.0
RULE 2	IF exactly 2 events occur within 2 instances at the input cheese – THEN an event occurs at the output allergy. S = 6; Q = 100.0 %, J^* = 87.6
RULE 3	IF exactly 2 events occur at the inputs juice and milk – THEN an event occurs at the output allergy. S = 4; Q = 100.0 %, J^* = 79.2
RULE 4	IF within 3 instances the pattern 1–0–1 occurs at the input cheese – THEN an event occurs at the output allergy. S = 3; Q = 100.0 %, J^* = 73.7

Table 2: The first four rules of the rule-based model in the illustrative example "Allergy". For each rule the situation number S, the quota of events Q and the relevance index J^* are noted.

For the specification of the statement space we assumed here that certain preknowledge about the process "allergy" is available. Otherwise – analogously to classical modelling – the complexity of the statement space can be increased stepwise until the model performance is satisfying.

3.2 Controller design example

In this section we illustrate how the above modelling concept can be applied in the field of control engineering. As a first example we improve the performance of a control system (fig.3): We model the "good decisions" of the (partly disturbed) actions of the two-level controller and thus obtain a rule-based controller which yields a better system performance than the original two-level controller.

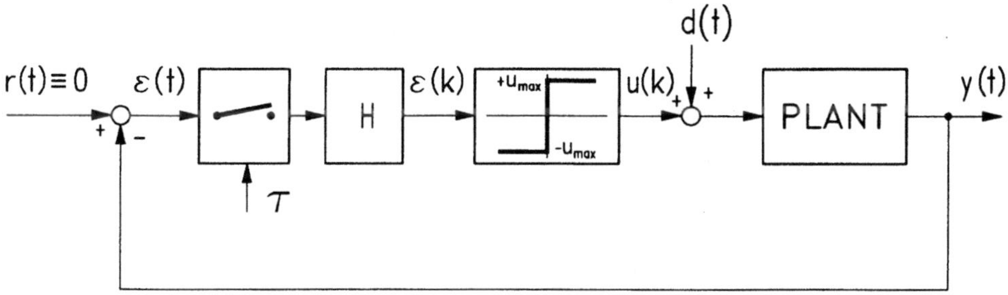

Figure 3: The performance of this control system with a linear plant, a two-level controller and a sample and hold unit was improved by the rule generating expert system ROSA.

First we embed the two-level controller, which obeys the two rules

$$\begin{array}{l} \text{IF } \varepsilon(k) \geq 0 \text{ THEN } u(k) = +u_{max} \\ \text{IF } \varepsilon(k) < 0 \text{ THEN } u(k) = -u_{max} \ , \end{array} \tag{10}$$

into a bigger space of qualitatively related statements, which additionally take into account the average of the deviation $\varepsilon(k)$ in the past: We define potentially relevant input events $e_1, e_2, \cdots, e_5$ by the decision functions

$$e_i(k) = \begin{cases} 1 \text{ if } \sum_{h=1}^{i} \varepsilon(k-h+1) \geq 0 \\ 0 \text{ else .} \end{cases} \tag{11}$$

For valuation of the controller decisions we declare each applied control input $u(k)$ of the operating system (fig.3) as "good", if the average of the future values $|\varepsilon(k+1)|$, $|\varepsilon(k+2)|$, $|\varepsilon(k+3)|$ and $|\varepsilon(k+4)|$ is smaller than a chosen limit ε_L. According to this we define the output event a by the decision function

$$a(k) = \begin{cases} 1 \text{ if } u(k) = +u_{max} \text{ is good} \\ 0 \text{ if } u(k) = -u_{max} \text{ is good} \\ ? \text{ if } u(k) \text{ is not good.} \end{cases} \tag{12}$$

The next step is to observe the control system and to transform the records of the sequences $\varepsilon(k)$ and $u(k)$ into a discrete event substitute system with the input sequences $e_i(k)$ and the output sequence $a(k)$ (according to fig.1). Finally these sequences are modelled using the expert system ROSA.

As a concrete example we chose a third-order continuous linear plant with the eigenvalues $\lambda_1 = 0$ and $\lambda_{2,3} = -0.1 \pm 2.34j$ and the sampling period $\tau = 1$. For the disturbance $d(t)$ we chose the step function $d_1(t)$ with the pattern shown in fig.4 and we specified $u_{max} = 1$ for the two-level

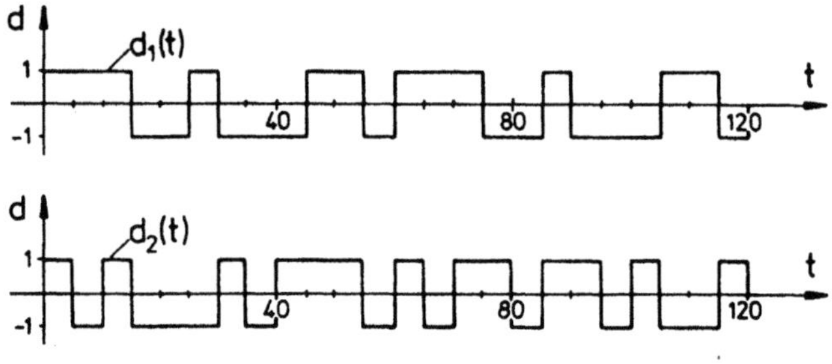

Figure 4: The disturbance $d_1(t)$ was applied in the learning phase for the rule generation, $d_2(t)$ was applied to test the obtained rule-based controller.

controller. As base for the rule generation ("learning phase") we disturbed the two-level controller: We stochastically selected 20% of the sampling instants and inverted the corresponding controller output. The underlying idea is that the introduction of random control decisions is necessary to find control rules which yield a better performance than the original two-level controller. This system was simulated up to $k\tau = 120$. For the limit ε_L, which enters the definition of $a(k)$, we took the value $\varepsilon_L = 0.5$ from inspection of the system behaviour operating with the original two-level controller. Analysing the resulting input and output sequences $e_i(k)$ and $a(k)$ by the expert system ROSA we obtained a set of rules which can be condensed (by inspection) into the following four control rules:

$$\text{RULE 1:} \quad \begin{array}{l} \text{IF } (e_1(k) \neq e_1(k-1)) \wedge (e_2(k) = 1) \wedge \\ [(e_3(k) = 1) \vee (e_4(k) = 1) \vee (e_5(k) = 1)] \\ \text{THEN } u(k) = +u_{max} \end{array} \tag{13}$$

$$\text{RULE 2:} \quad \begin{array}{l} \text{IF } (e_1(k) \neq e_1(k-1)) \wedge (e_2(k) = 0) \wedge \\ [(e_3(k) = 0) \vee (e_4(k) = 0) \vee (e_5(k) = 0)] \\ \text{THEN } u(k) = -u_{max} \end{array} \tag{14}$$

$$\text{RULE 3:} \quad \begin{array}{l} \text{IF (RULE 1 is not applicable)} \wedge \\ \text{(RULE 2 is not applicable)} \wedge \, (e_1(k) = 1) \\ \text{THEN } u(k) = +u_{max} \end{array} \tag{15}$$

$$\text{RULE 4:} \quad \begin{array}{l} \text{IF (RULE 1 is not applicable)} \wedge \\ \text{(RULE 2 is not applicable)} \wedge \, (e_1(k) = 0) \\ \text{THEN } u(k) = -u_{max} \end{array} \tag{16}$$

It is obvious that these four rules represent a rule-based controller which inverts the decision of the original two-level controller (10), if the values of ε in the past "strongly" say something different compared with the actual value $\varepsilon(k)$. This is qualitatively reasonable.

We investigated this "ROSA-controller" (13 - 16) quantitatively in different ways. Fig.5A shows the system performance – characterised by $\varepsilon(t)$ and $u(t)$ – for the original two-level controller and fig.5B for the

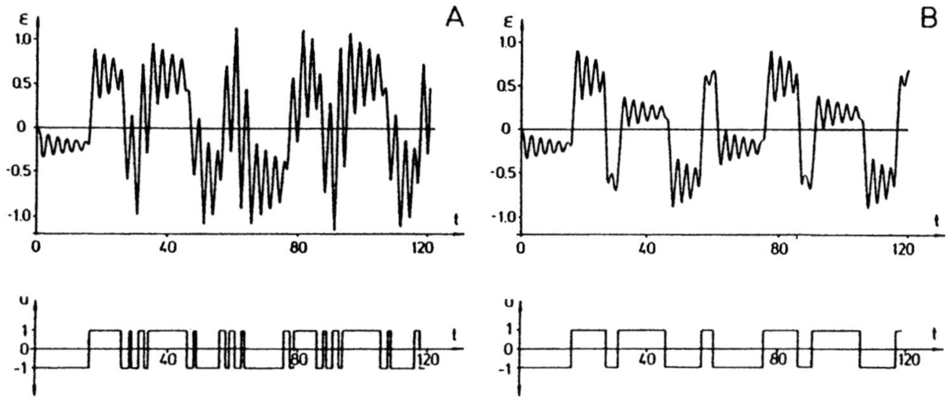

Figure 5: Control deviation $\varepsilon(t)$ and controller output $u(t)$ for the control system of fig.3 applying the disturbance $d_1(t)$ and using the original two-level controller (A). Replacing it by the ROSA-controller the performance is considerably improved (B).

ROSA-controller. In both cases the disturbance $d_1(t)$ was applied, which was used in the learning phase. We see that the ROSA-controller yields smaller control deviations $\varepsilon(k)$. Moreover it works with fewer change-over actions, which is plausible because a better controller hasn't to correct itself so often. Fig.6A shows the resulting frequency distributions of $\varepsilon(k)$ applying the random disturbance $d_2(t)$ shown in fig.4. Again the ROSA-controller (curve b) is better than the original two-level controller (curve a). This shows that the obtained rules represent basic regularities and not only accidental correlations in the learning sequences. The same shows up if we additionally change the plant parameters, so that the new eigenvalues are $\lambda_1 = 0$ and $\lambda_{2,3} = -0.08 \pm 1.79j$ which means a reduction of the corresponding natural frequency to 80% of its former value (fig.6B). The ROSA-controller (curve b) shows better robustness properties than the original two-level controller.

We established the quality of the ROSA-controller by simulation. In addition the rule-based controllers considered here allow for a great class of

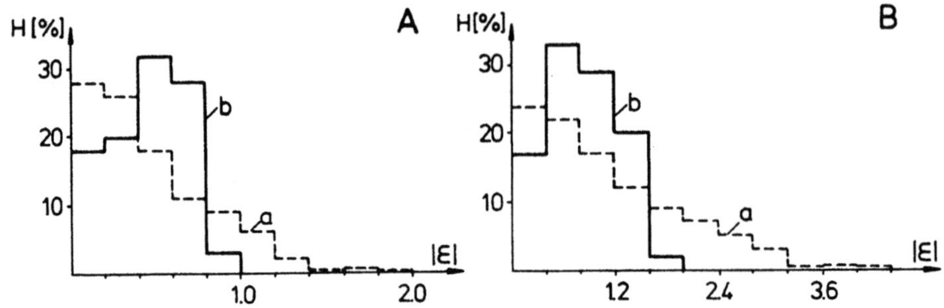

Figure 6: Frequency distributions of the absolut values $|\varepsilon|$ of the control deviation, expressed by the percentage H of the simulation time during which $|\varepsilon|$ lies within certain intervalls. The curves a correspond to the original two-level controller, curves b to the ROSA-controller. Fig.A shows that for the control of the plant which was used in the learning phase the rule-based controller is better than the original two-level controller. Fig.B illustrates, that the same holds if the plant parameters are changed.

plants to establish stability of the resulting control system: This is possible by the method of convex decomposition – introduced in [7] and further developed in [8] – if the plant is linear (as in this example) or additionally contains piecewise affine nonlinearities. Nevertheless we would like to stress that the method can also be applied, if no mathematical model of the plant is available.

4 View on further Applications

The above examples show that the presented method for generating rule-based models is a promising approach. Up to now we selected and started to prepare the following other applications:

- rule-based change-over switching between several controllers in a slowly time-variant system of which only the input and output variables are measured,
- rule-based selection of strategic elements and operations for the control of numeric parameter optimization procedures,
- rule-based controlling of discrete event processes (especially the generation of rule-based strategies for dispatching customers and servers in queueing problems).

These problems have in common that at certain instances decisions between a finite number of alternatives a_j (output events) have to be made. By the use of heuristic preknowledge observable indicators e_i can be specified, which describe the system state and are potentially relevant to control decisions. The system to be modelled is an unknown discrete controller for which the inputs are observed and good control outputs can be obtained from simulations. Thus in a direct way a rule-based controller is obtained by generalizing the observations of system behaviour.

Remark: This paper is a revision of the contribution "Regelorientierte Modellierung dynamischer Systeme mit Hilfe des hierzu entwickelten Expertensystems ROSA" of H.Kiendl and M.Krabs to the "10. Salzhausener Kolloquium der Automatisierungstechnik" in October 1988.

Acknowledgement: The authors would like to thank the "Deutsche Forschungsgemeinschaft" for the support of this work.

References

[1] Bieker, B.: Wissenserwerb für eine einfache Expertenregelung. Automatisierungstechnische Praxis 28 (1986), S. 448 - 457.

[2] Michalski, R. S., Carbonell, J. G. and Mitchel, T. M.: Machine Learning, An Artificial Intelligence Approach. Vol. I, Springer-Verlag, Berlin 1984, Vol. II, Tioga Publishing Company, Los Altos, California, 1986.

[3] Cohen, P. R. and Feigenbaum, E. A.: The Handbook of Artificial Intelligence. Vol. III, Tioga Publishing Company, Los Altos, California, 1982.

[4] Kiendl, H. and Krabs. M.: Ein Verfahren zur Generierung regelbasierter Modelle für dynamische Systeme. Zweiter Workshop über "Grundlagen für den Einsatz von Experten-systemen in der Automatisierungstechnik" 1988 in Hahnenklee. Abstract in Kiendl, H.: Zweiter Workshop über Grundlagen für den Einsatz von Experten-systemen in der Automatisierungstechnik. Automatisierungstechnik 37 (1989, S. 32 - 36).

[5] Kiendl, H.: Globale Parameteroptimierung unter Nutzung regelbasierter Strategieelemente. Dritter Workshop über "Expertensysteme in der Automatisierungstechnik" 1989 in Hahnenklee. An abstract will appear in Automatisierungstechnik.

[6] Kiendl, H. and Krabs, M.: Ein Verfahren zur Generierung regelbasierter Modelle für dynamische Systeme. Will appear in Automatisierungstechnik November 1989.

[7] Kiendl, H.: Robustheitsanalyse von Regelungssystemen mit der Methode der konvexen Zerlegung. Automatisierungstechnik 35 (1987), S. 192 - 202.

[8] Karweina, D.: Rechnergestützte Stabilitätsanalyse für nichtlineare zeitdiskrete Regelungssysteme, basierend auf der Methode der konvexen Zerlegung. VDI Forschungsberichte Reihe 8 Nr. 181, Düsseldorf 1989.

Multi-Sensory Signal Fusion

Armin Zimmermann

Abstract

This paper is to serve as an introduction to the multi-sensory signal fusion. It represents a literature survey and particularly shows the relations to Artificial Intelligence.

At the beginning a rough definition of the term fusion is given and the necessity of fusing sensor data is clarified. Later on the spectrum of the terminology is shown by explaining details of fusion and by remarking related problems. Finally the wide spread of applications and their relations to the field of Artificial Intelligence are enumerated and verified by various examples.

1 Introduction

1.1 What is Multi-Sensory Signal Fusion ?

First of all the expression "multi" has to be understood not primarily as a multiplication of one sensor but rather as **various** types of sensors. Because the special attention in this paper is directed to signals provided by (physical) sensors the expression "signal fusion" has been chosen ("sensor fusion" is used as well).

The third known expression "data fusion" is too general because it denotes also the fusion of data which are not provided by sensors. This can be for example intelligence data, like natural language narratives or descriptions. Of course these kind of data can be helpful to narrow a search-tree during the data-interpretation process, e.g. the recognition of a certain shape.

The fusion process is roughly the combination (integration, merging or pooling) of signal data provided by different sensors. Details on the steps to be done before and after fusion, related problems and how to separate fusion into sub-problems are given later.

1.2 Why is Fusion necessary ?

It can be stated that despite many years of intense research on the development of sensors and of computational theories for the interpretation of diverse sensor information, no single sensor can guarantee to deliver acceptably accurate information all the time (Brady, 1988).

One reason for this is the primitive stage of development of sensors for modalities as touch, proximity sensing and active ranging over distances of 5 to 10 m.

Another reason is the fact that a sensor is limited in its useful operating range. Interpreting data of a sensor driven beyond its bounds can easily lead to incorrect inferences. Examples concerning these limitations and misinterpretations are the saturation and hysteresis of touch sensors, the blasting of microphones and sonar[1]) specularities. Fortunately, sonar specularities can be identified in many cases, so the raw range data can be modified on the basis of the sonar sensor model.

Another important reason for inaccuracy of sensor information or of a signal is its inevitable corruption by noise, e.g. currently available touch sensors have poor signal-to-noise characteristics. Vision sensors require sufficient levels of illumination to be effective and a sonar signal transmitted through air attenuates strongly and is influenced by temperature and humidity.

So every sensor has its own characteristic strengths and weaknesses and therefore, it can be advantageous to combine data from differing types of sensors to eliminate their individual weaknesses and further enhance their strengths (Jollay, Ricks, 1988). For example, a sonar sensor and an infrared ranging device can be used profitably in combination. Sonar can give good range measurements for smooth portions of surfaces, but the angular directions of depth and surface discontinuities are often badly localized. However, even primitive infrared ranging devices can accurately detect edges or similar discontinuities, though the speed of light makes absolute ranging difficult. With these two sensors it is possible to construct maps of a robot's navigation room (Flynn, 1988).

Determining what combination of sensors to use is a function of many variables, e.g. the task and the environment of the sensor system, and the available computing resources.

So far the sensor fusion has been discussed due to technical limitations of sensors, but even when two or more sensors are operating within their limits, combining their measurements can provide a more **robust** or **reliable** reading than that provided by any one sensor. Sensor errors can be recovered and ambiguities can be eliminated.

For a complete image of the environment it is even necessary to fuse the signals of various sensors similar to the human sensory system, where

1) According to the literature the term sonar has to be understood in this paper as a rangefinding device which works with ultrasonic waves (here most often in air instead of water).

redundancy also is useful. In addition it can be more efficient to understand the environment or a certain scene by the use of easily acquired information from a number of sensor sources than can exhaustive analysis of the information from a single source.

For completeness it should be noted that it can of course be necessary to combine sensors producing the same type of data. Sensors working together with overlapping fields of view can generate hypotheses which may reinforce or conflict with each other during fusion. Another example is the usage of a number of the same sensors (e.g. cameras) for surveillance of a wide area in which one sensor is not sufficient (surveillance network) or for tracking tasks in an environment with hindering walls.

2 Kinds of Fusion and Related Problems

When browsing through the literature of fusion one finds many papers on the specific subject of "fusion" or even "sensor (signal) fusion". Unfortunately most of the papers only contain the description of the application and the used sensors, but only a few describe practicable methods how to fuse sensor information actually and which mathematical methods to use. Although this paper is not meant to be a recipe for fusion, it hopefully helps to clarify this subject, at least by pointing to the more helpful literature.

Closely related to the fusion of data (as well raw data as symbolized data) is the **representation** of sensor information, because different sensors provide different sorts of information (depth, surface orientation etc.). Some sensors provide dense samples of data (e.g. sonar) and others provide data only sparsely (touch). In addition it must be taken into account that different sensors take their measurements from different vantage points, so transformation methods are necessary before fusing. The last problem to be mentioned here (and there are still some more problems concerning fusion) is the question of how much processing can be done on the single sensor information before the final fusion process is carried out.

2.1 Categorization

Fusion can be categorized for example as either **local** or **global** and fusion is done on a more numeric level or a more symbolic level, where already processed or reduced data can be found. Local fusion means combining the signal data of sensors which are all collocated on a single platform. For the combination of data of widely distributed sensors the expression "global fusion" can be applied (LaJeunesse, 1986).

Categorization can also be done by following the phases of information acquisition. "**Far Away**" Sensing means the acquisition of information about a scene from a far distance. For instance, position, orientation, shape, size of objects, ambient temperature, and radiation levels are obtained by sensors which can perform adequately from a distance. During "**Near To**" Sensing it is not always possible to see the entire object in one view (no longer determination of shape or size), but rather to update e.g. position and orientation by proximity and visual sensors. "**Touch**" Sensing uses mainly tactile sensors for obtaining more detailed information of objects. During the "**Manipulating**" phase, the sensor system must keep track of all the information necessary to manipulate the object, and therefore sensors to detect force/torque, slippage, and weight are useful (Luo et al., 1988).

2.2 Chronologically Ordered Steps

The multi-sensory signal fusion can be seen as a collection of more or less separable sub-problems (Fernandez, 1985). It is embedded in a sequence of actions which have to take place on the way from the raw sensor signals to the displayed or monitored environment or a production process. The actions or tasks to be done range from collection, filtering (reduction of redundant information) and preprocessing of signal information via the fusion tasks like alignment, correlating and aggregating to the higher level tasks, as interpretation and displaying results (Wright, Gdowski, 1982).

The first fusion task is the time and space **alignment**. Relative sensor misalignments are the parallax error, base-to-base misalignment which is caused by the dynamic flexure and vibration of a sensor platform, and the line-of-sight-to-base misalignment which can be due to the environment, structural flexure and vibration (Bowman, 1979). The sensors work mostly asynchronous in time and space, hence interpolation of the samples and transformation of coordinates has to be done which is called asynchronous sensor fusion (Shafer et al., 1986). A simplified example is the merging of vision and rangefinder information provided by a sensor platform on a vehicle. The information referring to the camera or rangefinder coordinate system has to be converted to common coordinates, e.g. the vehicles coordinate system. Since the received data of either sensor have different timestamps, interpolation has to be done. The complexity of this step increases if every small deviation of speed and direction of the vehicle has to be taken into account.

The next sub-problem of the fusion process is called "**association**" or "**correlation**". New signal information sets, new images or features within it are correlated with previous data. LaJeunesse describes in 1986 the "first-level association" as registration of frames of data between different sensors or between sequences of frames from the same sensor. The correlation of features extracted from incoming data with features in the existing data base is called "higher-order association".

The correlation largely consists of **data base search** and comparison. Its goal is to determine if incoming information refer to entities already submitted, hence searching of an existing data base of entity information records has to be done. Sufficient effort has to be made to organize or to segment the data base into separate files with each file corresponding to a particular general entity type, e.g. all radar, infrared, and vision information would be placed in separate files. Additionally the entities in each file may be indexed by some key observable such as location. If incoming information describes an entity not present in the data base, then a new record containing all relevant information is created. If it is determined that incoming sensor information refers to an entity previously recorded, then the new and existing sensor information are merged to form one updated record containing the most accurate and timely status of the entity.

A helpful means during **comparison** is an association score computed for each record which has a reasonable chance of referring to the entity given in the incoming sensor information. The score will indicate the likelihood that the entity record and sensor information refer to the same entity. Well-known methods for approaching the signal data association problem as the Maximum-Likelihood/Hungarian Algorithm Methods, Procrustean Techniques, Sensor-Universe Mappings, Fuzzy Pointers and their usage in function of the circumstances are described by Fernandez (1985).

The merging or **combination** procedures for the updating of the records can be categorized into best information (the more specific information), weighted average (a procedure used on virtually all numeric data), and most recent observation.

The **aggregation** closely corresponds to the symbolic fusion. The received information is functionally grouped and organized by linking or relating lower level entities to higher level entities (parents). The other way round is necessary when higher level entities are observed which may be linked to their children already resident in the data base, since there can be no assurance that parents will always be observed before their children. The so-called "Clustering Aggregation" is the process of inferring the presence of high level entities based on information from a number of checked lower level entities (Wright, Gdowski, 1982).

2.3 Fusion Methods

Three methods for sensor fusion are described by Shafer et al. (1986). "**Competitive Fusion**" is typically used for sensors producing the same type of data, such as two sonar sensors working together with overlapping fields of view. Each sensor generates hypotheses which may reinforce or conflict with each other during fusion. "**Independent Fusion**" is described as a kind of degenerated method in which one sensor is used independently of another for each particular task. An important method is the "**Complementary Fusion**". Here the different sensors are used in a strategy that applies each one individually such that the relative differences of the sensors are used to enhance the advantages and cover the disadvantages of each.

Harmon et al. (1986) describe methods for the fusion of sensor data and the situations which lead to the selection of a specific fusion method. The simplest situation represents complementary sensor measurements. If two sensor observations are complementary and describe similar but independent situations then they can be logically added to the total environment description without concern for conflict. A more complicated situation arises when a dependency between measurements of different sensors exist (intersection of sensor observations, i.e. information of the same property value of the same object). The information must be merged by simple **interpolation**, if intersecting measurements represent observations at different times or spatial positions. Accuracies and confidences associated with interpolated values can also be interpolated.

The following three methods are fairly common if independent sensor measurements are completely intersecting (same object and property, same spatial location and approximately the same time). The **averaging** method uses confidence measures in the spirit that they are usually intended. This is a simple percentage calculation based on confidences which are then used as weights in an averaging of sensor data values. Given are values (V1 ... Vn) and their associated confidences (C1 ... Cn) from sensors (S1 ... Sn), then averaging weights can be computed by $Wi = Ci/(C1+...+Cn)$ and the fused estimate becomes $Vf = W1*V1+...+Wn*Vn$. The weight in the fused value can be most simply calculated from analogous equations $Wf = Vf/(V1+...+Vn)$ and the fused confidence becomes $Cf = W1*C1+...+Wn*Cn$.

Deciding is a fusion method which makes a discrete choice between several different measurements. For example the measurement with the highest confidence can be chosen or Bayes rule can be used to compute the resultant confidences of fused values. **Guiding** uses the values from one sensor to guide the control or processing (e.g. thresholding) of other more accurate estimates from another sensor.

2.4 Related Aspects

Probabilistic reasoning is fundamental to data fusion problems, since unreliable measurements are provided with accuracy and confidence factors. Methods such as Bayes theory or the Shafer-Dempster reasoning help in solving those problems (Garvey et al., 1981; Goodman, 1987). Bogler argues that the Shafer-Dempster reasoning is more applicable, if the information from disparate sensor sources with different levels of abstraction has to be combined.

Durrant-Whyte (1988) emphasizes the importance to provide an effective **model of sensor** capabilities. Henderson and Shilcrat introduce the idea of the "logical sensor", i.e. sensors can be defined abstractly in terms of computational processes operating on the output from other sensors (logical and physical). Advantages of this approach are the abstraction from implementation (only input/output relations or vectors are defined) and easy structuring, i.e. the logical sensor can be viewed as a network composed of subnetworks which are themselves logical sensors. With these features the developer of a logical sensor is insulated from peculiarities of physical sensors (range data can be produced by a laser rangefinder sensor or by two cameras and a stereo program) and can easily reconfigurate his sensor due to its modularization.

Moravec (1988), Elfes and Matthies (1987) describe **grid-based representations** of the environment (e.g. of a mobile robot). The cellular representation of Elfes and Matthies (1987) is called the "occupancy grid" (Moravec: certainty grid representation), where each cell in the map contains a probabilistic estimate of whether it is empty or occupied. A Bayesian estimation scheme is used for combination of information obtained from sonar and stereo vision.

3 Fusion and its Application Areas

The main application area to be elaborated here is Robotics and its autonomous mobile systems. **Mobile Robots** can be used for example for public safety applications (nuclear plants, etc.). Subsystems of mobile robots which heavily depend on fusion are robot navigation, motion control, obstacle avoidance etc. Different types of sensors and their fused information are also used in **Surveillance Systems** e.g. for airports, in anti-intrusion systems for nuclear plants, or for offshore oil fields monitoring, highly automated factories or chemical processes. **Manufacturing** tasks as automated assembly and sorting often need the combination of vision and touch sensors. **Defence Systems** and their subordinated systems as battlefield management, defensive fire control surveillance, target acquisition, automatic target recognition, weapons fire control or antisubmarine warfare highly depend on input of various sensors.

As an example for the first application area, "Mobile Robots", the Ground Surveillance Robot concept described by Harmon (1986) should be mentioned. Other appropriate examples are offered by Flynn (1988) who depicts the navigation facility of a sentry robot Robart II and by Jollay and Ricks (1988) introducing HERMIES-II, a research robot for hostile environments.

The Ground Surveillance Robot (GSR) is designed to transit from a given starting point to a given finishing point, without knowing the specifics of the intervening terrain. The GSR must therefore be able to construct a map of the crossed terrain and plan its route from this information alone. The sensors necessary for that complex task must perceive the vehicle's position, orientation, local obstacles (or more importantly, free space), and provide sufficient information about the distant surroundings.

For terrain mapping a laser rangefinder, a high resolution (i.e. 512 x 490) gray scale solid state camera and a low resolution colour camera are mounted on a three degree of freedom platform which can be elevated 1.5 m and rotated in azimuth approximately 360 degrees. Camera and laser must interact to build a terrain map by using the contrast map obtained from the camera output to guide the discrete sampling of the computer steered laser rangefinder. Region range properties can be merged with relative depth cues from camera imagery to construct the terrain map.

The vehicle attitude sensor subsystem monitors the relative vehicle position, roll, pitch and heading angles, forward speed and rotational velocity. For instance, the vehicle speed is derived from a true ground speed sensor using track speedometer (higher low speed accuracy) and doppler radar for high speed or when track slippage conditions may prevail. Relative vehicle position is gained from dead reckoning sensors or speed measurements combined with those of vehicle angular motion. Absolute position (not continuously available) is provided by satellite and landmark navigation systems.

Obstacle avoidance including obstacle mapping is handled by proximity sensors with some help from the mentioned gray scale camera. The proximity sensors consist of Polaroid acoustic ranging sensors (3 fixed, 4 mechanical steered). Besides finding a path through an obstacle field toward a distant goal, it is possible to follow a moving vehicle or to scan a wall. A local obstacle map is constructed from range and bearing information together with position information. Mainly averaging and guiding fusion methods are mainly used, i.e. averaging of data from the proximity sensor array and guiding of the camera steering through coarse proximity information.

The combination of sonar and infrared sensors for the navigation of the sentry robot Robart II is described by Flynn (1988). These two inexpensive sensors enable the robot to produce a refined map of its workplace and to plan a path through obstacles. The sonar is quite accurate in providing distance information but it has the effect of blurring obstacles and passageways, whereas the infrared sensor, although unable to provide accurate distance measurement, is able to detect obstacles or doors in walls with very good angular resolution. One simple heuristic rule for sensor fusion is worded as follows: "If the sonar reading is greater than the maximum range for the near-infrared sensor, then ignore the near-infrared output" (Flynn, 1988).

For its navigation and obtaining of 3D information the research robot HERMIES-II uses the combination of a single camera on-board the robot with sonar rangefinders (Jollay, Ricks, 1988). The camera was used as the primary sensor for detecting obstacles while planning navigation, and the sonar sensors are used to determine obstacle proximity while navigating.

Very useful for object recognition in a manipulation or assembly system is the combination of the complementary sensor system consisting of passive stereo vision and an active exploratory tactile sensor (Allen, 1988). The vision processing provides sparse 3D data about regions of interest that are then explored by the tactile sensor so that the 3D structure including curved surfaces, concavities and holes can be recognized.

Complex operations such as automated assembly and sorting, i.e. accommodation to changes in workpiece position/orientation are only possible by input from various sensors. Ren C. Luo et al. (1988) describe the development of an intelligent robot workstation capable of integrating data from overhead vision, eye-in-hand vision, proximity, tactile array, position, force/torque, cross-fire, overload, and slipsensing devices. The sensors can be grouped into four stages of information acquisition (from "Far Away" to "Manipulation" see chapter 2). More abstractly, fusion is seen as the merging of separate items into a single entity. This entity should be in the form of a template (data frame) of the current information acquisition phase. The template should be obtained by combining the appropriate sensor data templates into one representative template.

Defence systems and their subordinated systems already mentioned need information for example from forward-looking infrared sensor in case of darkness or from a laser system which can detect obstacles as small as wires during a flight. Navigation signals from a global positioning system satellite and inertial navigation sensors on-board update the location of the aircraft. For target detection infrared cameras (detecting e.g. running engines) and television cameras sensitive to low light levels are used (LaJeunesse, 1986). Multiple, common sensors such as a surveillance network of identical radars are used as well as multiple, different sensors such as sonar, electro-optical detectors, and seismometers (Waltz, Buede, 1986).

Defence systems often take intelligence data into account, e.g. narrative descriptions in natural language about the behavior of the enemy, probable location of enemy tanks and tank shapes. Expert systems with heuristic inference rules are often used for the combination of multisensor data, the classification and displaying of results, e.g. inferring target identification.

4 Relations between Fusion and Artificial Intelligence

Relations between fusion and Artificial Intelligence (AI) exist in many ways. A direct relation is the use of **Expert Systems** (ES) for fusion, since many non-deterministic decisions have to be made. Even if the fusion process is not planned in form of an ES, several AI-techniques are used such as blackboard architectures for inference control, representation schemes for a consistent representation of sensor information (e.g. entity-property-value), heuristics formulated as rules, and inference methods such as hypothesize-and-test.

The strong relation between fusion and AI is exemplified by numerous examples. A real-time ES and a LISP processor are used in the HERMIES-II project (Jollay, Ricks, 1988), a rule-based system with a hypothesize-and-test method for object recognition (Allen, 1988), and ES inside defence systems for combining multisensor data and inferring higher levels of information about targets etc. (Waltz, Buede, 1986).

Reiner presented in 1985 a LISP-based ES. Typically it is able to handle non-deterministic decisions which can be expressed in terms of probabilistic/decision logic trees as used for problems concerning data association and fusion of classification declarations. Ambiguities during association, correlation and classification which frequently arise in multisensor estimation systems are resolved by expert rules (of thumb) and a hypothesis generation and test method.

LeClair (1986) describes a sensor fusion system for processing of advanced and emerging materials (here graphite epoxy laminates). The heart of the system consists of three expert systems programmed in FORTH. The interface or blackboard ES provides the file management tool and communications between the sensor interface, the control interface, the other two ES, and the user at the terminal. The sensor parser ES transforms sensed data into memory structures useful for planning, control and diagnostic purposes. The parser applies knowledge of the task domain to prevent ambiguities and to select the correct semantic meaning of the input data. The third ES, the analyzer has to execute user-supplied plans in order to control a task environment and it must be able to detect, isolate and resolve faults. The analyzer uses input such as user commands and task environment information posted on the blackboard.

An expert system for the fusion of radar and jammer strobe information is presented by Brogi et al (1988). This ES merges the sensor information with a priori known information to classify and possibly identify targets in a region under surveillance. A set of knowledge sources each containing the a priori expert knowledge about radar target classification cooperate to find a global solution. The partial solutions of each module which are represented by probability distribution functions (mass) have to be handled by the probabilistic reasoning method called the Dempster-Shafer theory of evidence. Each knowledge source is allowed to add/delete hypotheses to/from the blackboard which represents the only communication means for the knowledge sources and therefore the basis of the ES.

The **blackboard architecture** paradigm is often used in the sensor fusion domain, e.g. for the Ground Surveillance Robot and the mobile robot vehicle NAVLAB described by Harmon (1986) and Shafer (1986) respectively. However, the blackboard architecture is modified in that it has a distributed structure, since it handles data of distributed sensor subsystems. New features of traditional blackboards are necessary for sensory signal fusion such as hierarchical structure, parallel execution of modules, synchronization and geometric reasoning. The blackboard additionally is a conceptual device which provides a clear and consistent representation of information and blackboard instances respectively, e.g. in form of object, property, value, confidence, timestamp tuples. Inheritance properties of hierarchically organized classes enable an economy of representation and generalization.

Pau (1989) especially investigated **knowledge representation** approaches in sensor fusion. Since the information types of different sensors differ through physical sensing principles, data rates, bandwidth etc. a general knowledge representation scheme applicable to sensor fusion has to be formalized. The selection and structuring of the knowledge representation is highly critical for the performance of a sensor fusion system. The presented scheme is hierarchically decomposed, i.e. correlating, matching, aggregating, and projecting of sensor data icons is done step by step.

5 Concluding Remarks

The increasing interest of universities, industry and military institutes in researching the fusion of multi-sensory signals is derived from demands in highly automated manufacturing, modern combat etc. Mobile robots are needed to execute tasks under hostile conditions (deep ocean environments, outer space etc.). To perform complex tasks autonomously, mobile robots heavily depend on various sensors, computational models, and information processing methods such as fusion for the perception and interpretation of the surroundings.

It is shown that the combination of AI-techniques/Expert Systems and fusion systems enhance the capabilities of autonomous systems. Remarkably often AI-means such as the blackboard paradigm or a hypothesize-and-test method are used. However, the architecture of the traditional blackboard has to be extended due to the structure of sensor systems (i.e. hierarchical, distributed, parallel).

References:

Allen, P.K., Integrating Vision and Touch for Object Recognition Tasks, The International Journal of Robotics Research, pp. 15-33, 1988.

Bogler, P.L., Shafer-Dempster Reasoning with Applications to Multisensor Target Identification Systems, IEEE Transactions on Systems, Man, and Cybernetics, SMC-17, No. 6, 968-977, 1987.

Bowman, C.L., Multisensor Integration for Defensive Fire Control Surveillance, IEEE Proceedings of the National Aerospace Electronics Conference, pp. 176-184, 1979.

Brady, J.M., Foreword: Special Issue on Sensor Data Fusion, The International Journal of Robotics Research, pp. 2-4, 1988.

Brogi, A., et al., An Expert System for Data Fusion Based on a Blackboard Architecture, 8th International Workshop on Expert Systems and Their Applications, pp. 147-165, Avignon 1988.

Chair, Z., Varshney, P.K., Optimal Data Fusion in Multiple Sensor Detection Systems, IEEE Transactions on Aerospace and Electronic Systems, AES-22, No. 1, 98-101, 1986.

Durrant-Whyte, H.F., Sensor Models and Multisensor Integration, The International Journal of Robotics Research, pp. 97-113, 1988.

Elfes, A., Matthies, L., Sensor Integration for Robot Navigation: Combining Sonar and Stereo Range Data in a Grid-Based Representation, Proceedings of the 26th Conference on Decision and Control, pp. 1802-1807, Los Angeles, 1987.

Fernandez, M.F., Data Association Techniques for Multisensor Fusion, IEEE Workshop on Languages for Automation: Cognitive Aspects in Information Processing, pp. 277-281, 1985.

Flynn, A.M., Combining Sonar and Infrared Sensors for Mobile Robot Navigation, The International Journal of Robotics Research, pp. 5-14, 1988.

Garvey, T.G., et al., An Inference Technique for Integrating Knowledge from Disparate Sources, Proceedings of the 7th International Joint Conference of Artificial Intelligence, pp. 319-325, 1981.

Goodman, I.R., A General Theory for the Fusion of Data, Tri-Service Data Fusion Symposium, Laurel, 1987.

Harmon, S.Y., et al., Sensor Data Fusion through a Distributed Blackboard, Proceedings of the IEEE International Conference on Robotics and Automation, pp. 1449-1454, 1986.

Harmon, S.Y., Practical Implementation of Autonomous Systems: Problems and Solutions, Intelligent Autonomous Systems, An International Conference, pp. 47-49, Amsterdam, 1986.

Henderson, T.C., et al., MKS: A Multisensor Kernel System, IEEE Transactions on Systems, Man, and Cybernetics, SMC-14, No. 5, 784-791, 1984.

Henderson, T.C., Wu So Fai, A Multi-Sensor Integration and Data Acquisition System, Proceedings of the IEEE Conference on Computer Vision and Pattern Recognition, pp. 274-279, 1983.

Henderson, T., Shilcrat, E., Logical Sensor Systems, Journal of Robotic Systems, pp. 169-193, John Wiley & Sons, Inc., 1984.

Jollay, D.M., Ricks, R.E., Sensor Fusion for Robot Navigation, Oak Ridge National Lab., Technical Note, 1988.
LaJeunesse, T.J., Mission Success in Future Tactical Systems Requires Sensor Fusion, DS&E, pp. 21-31, 1986.
LeClair, S.R., Sensor Fusion: The Application of Artificial Intelligence Technology to Process Control, Proceedings of the 1986 Rochester Forth Conference, pp. 125-132, 1986.
Luo, R.C., et al., Dynamic Multi-Sensor Data Fusion System for Intelligent Robots, IEEE Journal of Robotics and Automation, Vol. 4, No. 4, 386-396, 1988.
Moravec, H.P., Sensor Fusion in Certainty Grids for Mobile Robots, AI Magazine, pp. 61-74, Summer 1988.
Pau, L., Knowledge Representation Approaches in Sensor Fusion, Automatica, Vol. 25, No. 2, 207-214, 1989.
Reiner, J., Application of Expert Systems to Sensor Fusion, Proceedings of the IEEE 1985 National Aerospace and Electronics Conference, pp. 1444-1450, 1985.
Richardson, J.M., Marsh, K.A., Fusion of Multisensor Data, The International Journal of Robotics Research, pp. 78-96, 1988.
Shafer, S.A., et al., An Architecture for Sensor Fusion in a Mobile Robot, Proceedings of the IEEE International Conference on Robotics and Automation, pp. 2002-2011, 1986.
Shekhar, S., et al., Sensor Fusion and Object Localization, Proceedings 1986 IEEE International Conference on Robotics and Automation, pp. 1623-1628, 1986.
Tong, R.M., et al., Distributed Hypothesis Formation in Sensor Fusion Systems, Proceedings of the 20th IEEE Conference on Decision and Control including the Symposium on Adaptive Processes, pp. 1421-1424, 1981.
Waltz, E.L., Buede, D.M., Data Fusion and Decision Support for Command and Control, IEEE Transactions on Systems, Man, and Cybernetics, SMC-16, No. 6, 865-879, 1986.
Woolett, J.F., MMWR/FLIR/ATR Sensor Fusion Proof of Concept, IEEE AES Magazine, pp. 22-25, 1988.
Wright, F.L., Gdowski, S., The Art of Multisensor Fusion and Correlation in a Tactical Environment, Proceedings of the IEEE Region 5 Conference and Exposition, pp. 90-93, 1982.
Yager, R.R., New directions in Multi-Sensor Fusion, Proceedings of the SPIE, The International Society for Optical Engineering, pp. 405-413, 1989.

6
Intelligent Robots

Goal Oriented Behaviour of Robots for Space Applications

O. El Zubi, R. Münstermann

Abstract

Within future space activities autonomous robots will have a broad field of applications. As an example they will be used for the exchange of defect ORU's (Orbit Replaceable Units) of satellites. In order to solve those tasks, plans must automatically be generated and executeded by using the methods of artificial intelligence. The planning and performance system KOMPLEX as described hereafter demonstrates the possibility to use a normal industrial robot as an autonomous system. This means, that the total robot movement will not be explicitely described, but will be generated by the planning system according to a goal setting generating an action plan and the execution system turns this plan into a sequence, which can be taken by the normal robot control unit.

Introduction

The robot technology is presently in a transition phase from the second to the third generation. This trend is heavily influenced by the discussion about manned or unmanned space activities. In the US as well as in Europe studies have shown that long range and long duration space enterprises cannot be performed success- fully without highly development automation and robot systems. Furtheron methods of artificial intelligence (AI) will play a dominant role for the realisation of those space programmes emphazising problem solving and knowledge representation.

The industrial use of robots started in the beginning of the sixties with robots of the first generation. The demand for a flexible machine which can perform many tasks with the help of stored programs was only achievable with the combination of the mechanical part and the computer technology. The flexibility was therefore achieved with the possibility of programming new sequences.

Robots of the first generation are characterized mainly by its mechanical design whereas the computer just functions as a storage place for these operations initiated by men. The efficiency of this generation is limited to stereotype tasks as the spot welding. This capability however, is just not enough for complex handling processes like arc welding. For this task complex technical provisions are required like trajectory controlling or sensor integration.

In order to solve complex tasks the programming itself became more and more important and defined the transition to the second generation. One important characteristic is that the robot languages for this generation are normally developed in a "standard" high-level-language, for example AL based on ALGOL. The features of these languages are structured programming, trajectory steering, transformation of coordinates, interrupt processing etc.

In order to comply with the common request, that robots must fullfill tasks outside of the industrial application, they must be equipped with "intelligence" leading to robots of the third generation. They must be able to define their own goals, to plan their own actions and must accomodate changes in their environment. Examples for these applications are fire fighting, deep water research and the underwater mining activities.

For this type of application there exist in general no algorithmic solution because these are undetermined and ambiguous symbolic processes. This forced scientists to develop new procedures. Therefore it was necessary to provide many solutions, which than can be tested. This new start of the explorative programming was the first step into the new research area of "Artificial Intelligence". This new and promising approach encountered however very soon many difficulties with complex systems. The number of possible solutions was explosively increasing and it was necessary to apply methods based on (heuristic) knowledge in order to reduce the amount of search drastically.

In general, AI leans on:

- o the knowledge about the field, which is of interest
- o the methods to use that knowledge and upon
- o the control structures to influence the process.

Description of the planning- and execution system KOMPLEX

The automatic plan generation for robot applications based on AI-methods is the major theme of the KOMPLEX-system described hereafter.

Within the past 20 years several planning methods have been developed: the hierarchical planning, the non-hierarchical and the skeleton planning.

The hierarchical planning in contrast to the non-hierarchical differs between important and less important considerations. Both procedures are supported during the planning by the description of the environment and by the research. The skeleton planning is in addition being supported by the experience knowledge. The characterization for these methods are, that they are appropriate for different knowledge basis but not in the same way.

The efficiency of the planning is dependent from the largeness of a domaine and from the planning method choosen. At realistic but complex heterogeneous problems, like robot applications the efficiency would rapidly decrease.

These reasons were decisive upon the development of a new concept: "Individual planning through Meta knowledge". This concept is based upon the fact, that the real and complex heterogeneous problems are composed of may added subproblems.

The individual planning is a method, by which the problem is not considered as the whole, but it will be hierarchical separated according to the degree of structure of the individual problems. According to each individual problem the most appropriate planning method is applied. It must be however considered, that the separated individual problems can be dependent from each other and therefore be separated before the real planning.

The realisation of the concept can be achieved through the controlling of the domain specific knowledge by the Meta-knowledge, which is superior. This Meta-knowledge ist basically:

- o the separation of the problem in problem shares through the separation rules.
- o the deconcentration of the problem shares with the help of the validitation conditions,
- o the individual treatment of each problem with the help of selection rules related to the method, the knowledge base and the goal and
- o the combination of shares of elemantary plans.

The realisation of this concept is the planning- and action system KOMPLEX, which was implemented with the AI-language PROLOG.

The modular build up of this system and the domain independent description of the rules allows the addition and the treatment of applied-specific knowledge of different domaines. The complete efficiency of the system was proofen when several application domaines were choosen. The planning- and execution system KOMPLEX demonstrates the possibility, to prosecute an ordinary industrial robot as an autonomous system. Autonomous means, that the whole movement is not explicitely being programmed, but the goals are only formulated. Example: put block a on block b (put on (a,b). In other words, the planning system is generating such a goal from an action plan and the execution system is changing this plan into a code which the robot control unit understands.

Example: The "ORU"-exchange problem

The formulation of the problem was created from a service task in orbit related to EURECA (european retrievable carrier) and HERMES, the european space plane. For both vehicles ORU's should be exchanged against reserve units. ORU's which are not needed should be put away beforehand (see Fig. 1). The exchange itself should be performed by a robot in application of AI-methods.

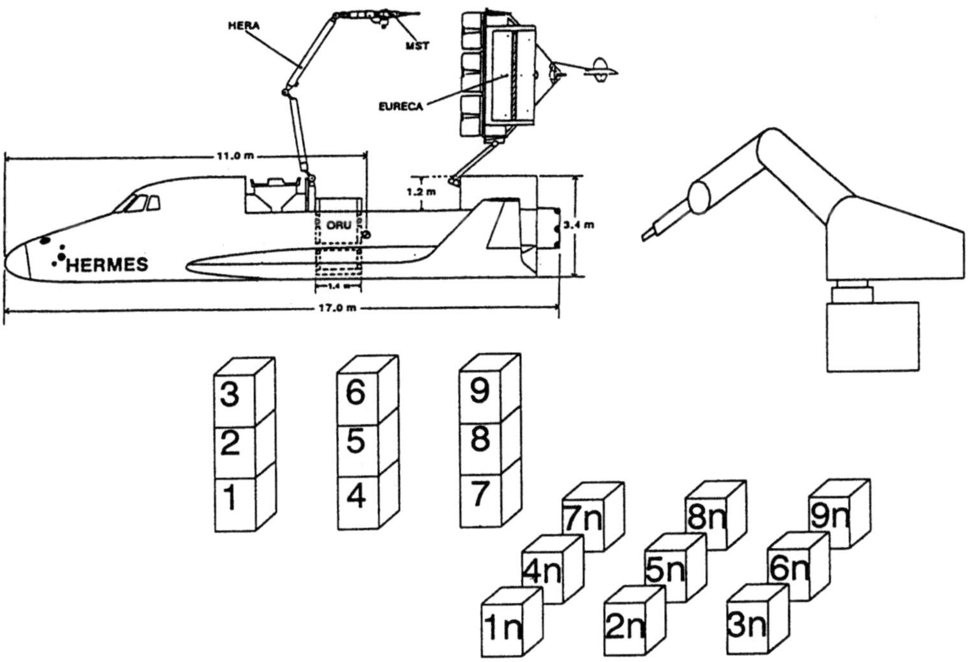

Fig. 1: task scenario

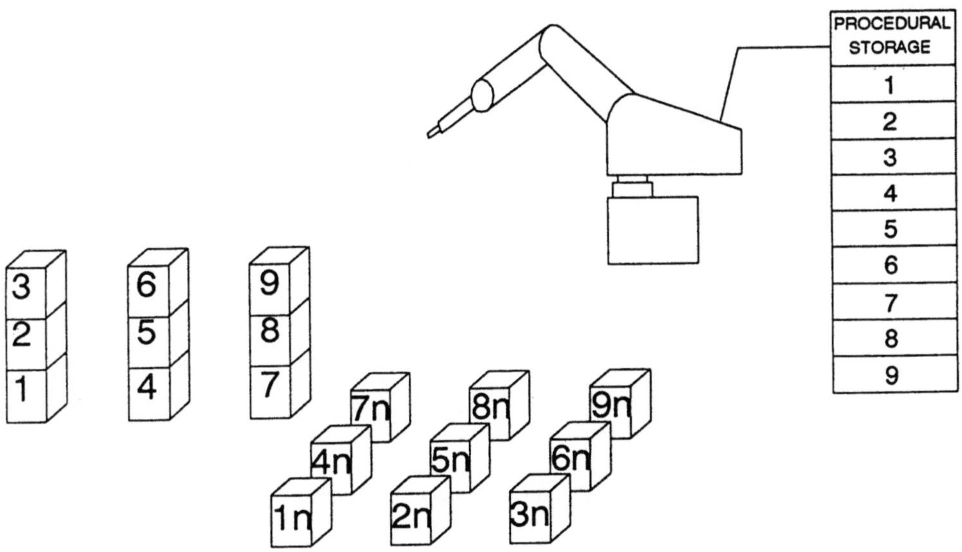

Fig. 2: procedral task solution

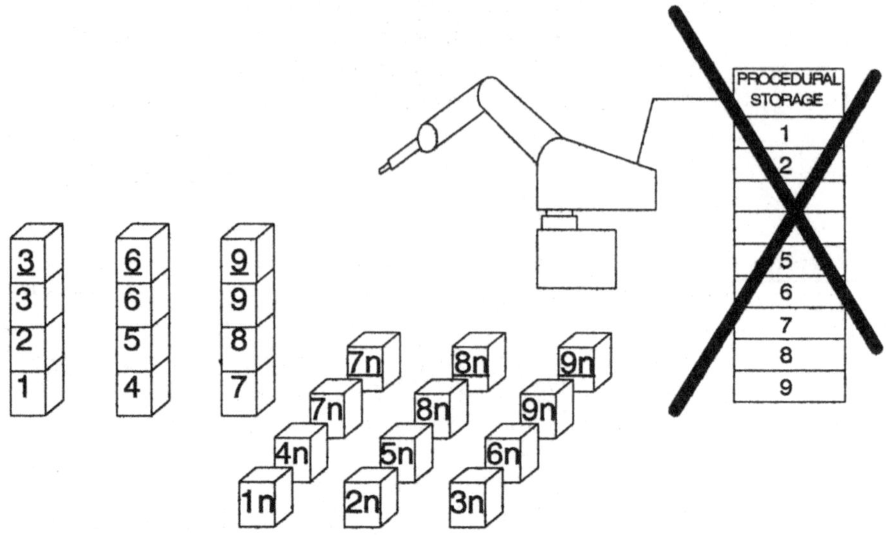

Fig. 3: fail of procedural solution

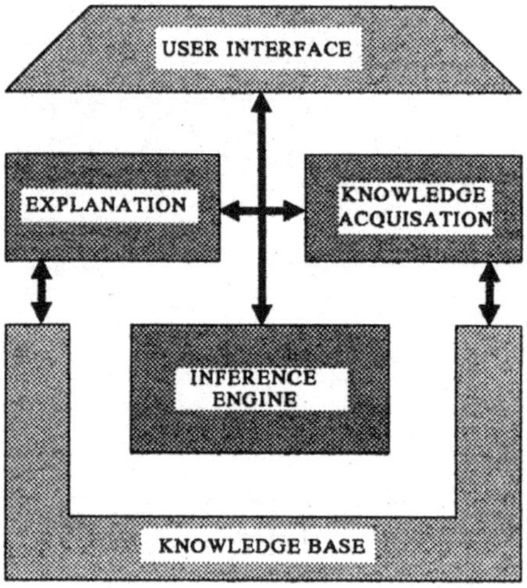

Fig. 4: expertsystem structure

The conventional method is the algorithmic solution of the task (Fig. 2). For the exchange process of each single block the piling procedure is programmed before and is stored accordingly. For example, should ORU 4 be exchanged, the routine no. 4 is recalled out of the store and the robot is working accordingly. The disadvantage of this procedure is, that if the starting point changes (Fig. 3), all other routines in the store become unuseable and new exchange procedures incl. the trajectory guidance of the robot have to be established again. Therefore, the solution with algorithm is not appropriate for this purpose. The alternative is the knowledgebase solution (Expertsystem Fig. 4). In this case the solution of the task will be performed via knowledge in the way of de facto knowledge and rule knowledge laid down in a knowledge base.

So called inference mechanisms are developing on the base of the stored knowledge the necessary sequences for (robot actions) the required tasks. At the same time the de facto knowledge and the rule knowledge can be varied via the dialogue element.

Basic Principles of Intelligent Task Planning for Autonomous Robot Systems

I. Lorenz, C. Massau

Abstract

Autonomous robot systems will be of major importance in the future space scenarios. New technologies have to be applied to increase autonomous intelligent task execution. This paper discusses some basic techniques to allow and to improve the flexibility and efficiency of automatic situational adequate task sequence generation, i.e. intelligent task planning. Based on some requirements for intelligent task planning, techniques of artificial intelligence fullfilling the requirements are discussed. The implementation of these techniques in two planning systems, developed at the University of Bremen under cooperation with the University of Oldenburg are presented.

1. Introduction

The european space scenarios as provided by the European Space Agency for the next thirty years, have lead to a discussion about operational costs of manned and unmanned autonomous space bases. As a result a partial manned base, the COLUMBUS Free Flying Laboratory, is under development. Operational analyses have shown, that the Free Flyer is only partial controllable from earth. Having Automation & Robotic (A&R) Systems onboard, the basic question arises how to ensure proper operation of the A&R System under all uncontrollable conditions. The most interesting solution path for this problem is to add "intelligence" to the onboard A&R Systems.

In the following sections, this theme will be discussed in a more principle manner. After a rough definition of levels of autonomy for robot system, requirements for an improvement of intelligence are established. The following section discusses some AI techniques, which are of basically importance for the requirements fullfillment. On this baseline two planning systems are designed and implemented dealing with special aspects fo the requirements set. As a conclusion, related work and some outstanding problems will be named.

2. Autonomous Robot Systems

2.1 Definition

An autonomous robot system can be defined as a system capable to try to reach an (intentional) user defined goal by selecting and executing appropriate tasks in a physical environment.

2.2 Levels of Autonomy during Task Planning

The highest level of autonomy can be assigned to a robot system which can act in a dynamically changing environment including the possibility of unpredictable goal affections. A rough functional description of such a system can be given as follows.

On the baseline of an initial environment description a complete plan for reaching the given goal is produced. This plan serves as a reference for the plan execution module and a robot internal environment simulator. During execution of a plan segment, an internal expectation about system state and environment changes is produced and in a second step compared with the actual sensor measurements. As a result, three lines of internal action can be taken

- execution of the next reference plan element in the case of measurements and expectation coverage
- goal consistency check and goal redirection if discrepancies between the resources necessary to reach the predefined goal and the actual state arises
- environment check and simulator update, if discrepancies in actual measurements and expectations are detected.

The both latter cases lead to a replanning process either to reach the redirected predefined goal or planning of recovery tasks due to environment changes.

A lower level of autonomy is possible for robot systems if it is possible to guarantee that the user defined goal can be reached under all circumstances. The complex process of goal redirection and consistency check can be eliminated.

The lowest level of autonomy is possible for robot systems which are acting in a static (i.e. closed) world. This means, that the system has no ability to recognize changes in its environment which are evoked from outside. The robot system produces a plan of tasks on the initial environment configuration baseline to reach the predefined goal, executes "blind" all tasks and changes its internal environment model in accordance with the plan execution. Well known example for these conditions is the blocksworld, developed and described by Winograd /WINO/.

It is an already complex task to reach this lowest autonomy level. In the following sections, some requirements and problem examples for a robot task planning system are elaborated and possible basic solutions based on artificial intelligence techniques are discussed in the scope of the blocksworld-paradigm.

3. Requirements for a Task Planning System in the Blocksworld Paradigm

Major aim of this section is to introduce some criteria on which the degree of "intelligence" of a robot system can be evaluated. On the other hand, these criteria are requirements for an "intelligent" design of a robot system's task planning module.

- o The planning system has to be flexible.
  The planning system must be able to construct solutions not only for a special problem but also for the entire class of similar problems.

- o The planning system has to be goal directed.
  The planning system must be able to reach a specified goal from an arbitrary initial state without further external advise.

- o The planning system has to be optimal.
  This requirement can be devided into completeness and optimality due to a criterion.

  - The planning system has to be complete /CHAP/.
    Within the choosen environment (i.e. Blocksworld) the planning system has to produce a solution under all allowed circumstances.
  - The planning system has to be optimal due an optimality criterion.
    The planning process must be able to minimize a given problem related criterion.

- o The planning system must be able to communicate with the user.

There are still some more requirements, but these seem to be the most important one's dealing especially with the "intelligence" of a planning system. The next two sections will first explain some possible ideas based on artificial intelligence techniques for covering the requirements mentioned above and will secondly present two plannig system implementations dealing with special aspects of the requirements.

4. Applicable basic AI Techniques for Planning Systems

Main issue to reach the required flexibility for a planning system is the separation of the planning procedure in three separate parts:

- the world model (i.e. the environment)
- the operators which act upon the world model
- the procedure which finds an appropriate operator.

The world model can be seen as a set of relational entities. It describes the objects of the (closed) world, their physical properties (magnitude, weight, ...) and their relationship to other objects of the world. The types of relationships and the properties are assumed to be unchangeable.

The only possibility to change the state of the world is the application of an operator, which can be seen as a task performed by the robot system. This means, that the meaning of an operator is twofold:

- on one hand is the description of the task to be performed by the robot system
- on the other hand lies the applicability, defined by the state of the world which has to be reached before application, and the affected state elements after operator application.

So an operator is described by

- the task to be performed
- the application preconditions
- the post conditions.

From this, it can be seen, that the separation concept described above guarantees the required flexibity. The operator application search procedure can be easily designed to check the preconditions of all operators to find a situational appropriate operator. So, if the planning system is complete and goal directed, the class of problems defined by the operators can be solved.

To reach the required goal direction, the AI community provides a famous algorithm, known as MEA (Means-Ends-Analysis) algorithm /BAF/. In principle, the MEA-algorithm is bound to the operator structure above and represents the basic procedure to find an appropriate operator.

The MEA-algorithm works as follows:
starting with the user goal, an operator will be selected which contains in its postconditions the state element, the user wants to be changed. In a second step, the preconditions of the candidate operator will be checked against the actual world state. The third step decides wether the operator is applicable (i.e. all preconditions are met by the actual state) or which preconditions have to be established as subgoals. Step four either finished the MEA algorithm because the user goal or a subgoal is reached or decides which subgoal has to be solved by a recursive call of the MEA-algorithm. The MEA algorithm guarantees that operators are selected, which connect directly the goal with the actual initial state of the world.

To reach optimality in planning systems is a hard, if not an impossible task. Most crucial is the completeness of a planning system. Completeness means, that an arbitrary number of user goals can be fullfilled by the planning system simultaneously. There is no problem to solve one user goal at a time within the MEA algorithm. It is the treatment of conjuctive goals that is more problematic, where more than one user goal has to be reached during one planning process. A simple example is the so called Sussmann-anomaly, depicted in figure 1 /WAL/.

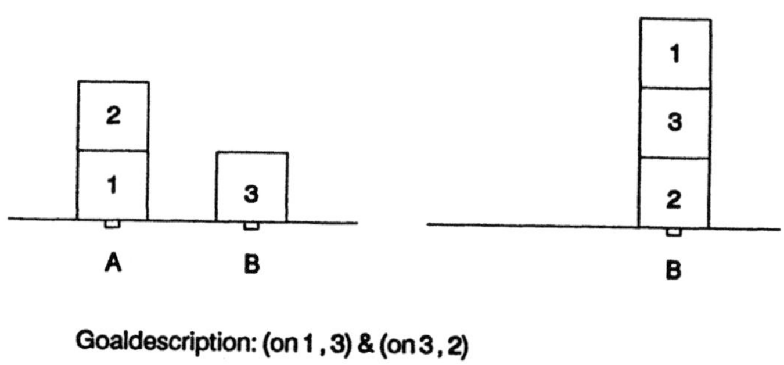

Fig. 1: Sussmann Anomaly

Sequencing the goals leads to a destruction of the already reached goal. Thus in prinicple, the MEA algorithm itself is not able to solve this problem because there is no recursion over more than one user goal.

Possible solutions of this problems are
- goal regression, developed by Warren and Waldinger /WAL/
- search in the space of possible plans /LOM/.

Goal regression works with protected states reached by solving a user goal. The protected state consists of the set of objects and their interrelationships. This unchangeable closure can be handled as a new (intermediate) object with new properties. These are defined by the relationships pointing outside of the protected closure. So the second (and so on) goal can be solved without closure destruction using the external intermediate closure properties. It should be noted that the principle of goal regression has no physical equivalent. This is worth to be mentioned in terms of realtime planning and plan execution, because by principle the closure cannot be broken.

Searching in the space of possible plans deals more with optimality as with completeness, but the results are in principle the same. The search method developes a tree of possible actions due to the initial state. Every branch to a node (possible state) has a weighting factor caused for example by the number of actions to reach the assigned state.

The next step after tree development is state evaluation of the nodes with respect to the desired goal state. If the goal state is reached by one node, the total weighting factor is added and all other branches are stepwise developed as long as their weighting factor is greater or equal to that of the solution node. This guarantees that no better solution due to the action optimality criterion exists. This procedure, known as Branch-and-Bound search /WIN1/ method allows the combination of the aim of reaching completeness and optimality due to minimize an appropriate criterion.

In comparison with the goal regression procedure, branch-and-bound is also a realtime planning critical procedure.

However, the branch-and-bound method can be viewed as a "look-ahead-facility" in a realtime planning system because every step in the process of possible solution tree development can be realized physically in contradiction to the nonphysical closure construct during goal regression.

Fullfilling the communication requirement, two main aspects have to be handled
- the system must be able to explain its behaviour and the relationship between the objects in its world model
- the system must be able to accept and validate user commands.

The first aspect deals with the well known explanation facility in the expert system concept. Classical procedures during the explanation process are the answering of "why", "how" and "when" questions. Like in the classical expert system approaches of explanation by preformulated text or instantiated text patterns and rule firing history, an answer can be constructed by storing and reviewing the solution tree of the operator/world state sequences.

The procedures in short are:
- a "why" question is answered by the parent node of the instantiated question pattern,
- a "how" question is answered by the child nodes of the instantiated question pattern,
- the "when" question is answered by the pattern associated to the top node (user goal).

The second aspect of communication deals with restriction of user input. It is necessary to control user inputs in terms of a valid command syntax and in terms of command meaning, i.e. and unambigous projection of user command to planning system procedures. This specification is supported by techniques known as augmented transition network (ATN) /WIN1/ parsing. Baseline is a restricted set of syntax patterns which define the set of sentences and terminals the system accept. The semantic of known system commands is placed in a lexicon which associates symbols with proporties and procedures. The parsing process itself is a depth first search and match procedure.

To analyse and to implement the techniques mentioned above, two planning systems were developed, each with a special scope
- the first system, MINOPTEROPS (multiple independent goals nonoptimal operator sequence planning system) takes special emphasis on the elaboration of flexibility and goal direction
- the second system, COLAPS (conjunctive goals and optimal operator sequence planning system) deals especially besides the first two aspects with optimality and user communication.

A short description of both systems is given in the next section.

5. Description of the implemented Prototypes

5.1 MINOPTEROPS

MINOPTEROPS /LOM/ stands in the historical line of the STRIPS planning systems, first developed by Fikes and Nilsson/FIN/. The characteristics of this family are

- separate relational world model
- operator definition by naming operator task, application preconditions and application post conditions
- MEA algorithm for search of applicable operators.

From this, the special implementation features of MINOPTEROPS are

- the flexibility principle
- operators and world model for the blocksworld
- a complete MEA-algorithm
- autonomy on the lowest level.

An easy way to achieve a basic flexibility is the application of a rule based implementation language. By this, the application conditions of each step during the work of the planning system can be explicitly controlled. So a clear separation of rules belonging either to the MEA algorithm, to operators or to the world model is possible. As implementation language for MINOPTEROPS , OPS5 /BRO/ was choosen. OPS5 is a typical production system language, controlled by the well known recognize-act cycle. The user accessible parts of OPS5, i.e. a rule memory and a separated working memory, support the flexibility requirement. The world model as the changing part in a planning system is implemented in the working memory, the operators and the MEA algorithm as the static part are expressed totally in the rule memory. It should be noted here, that OPS5 is not an ideal implementation language because an implementation of the goal directed recursive MEA algorithm is very inefficient in a language providing only "forward" (i.e. situation directed) chaining. But as mentioned above, direct control about variable binding, state of recursion and goal satisfaction can be taken by examining the work memory after each rule application.

Special emphasis during design and implementation of the operators and the world model was taken on the so called frame problem. This term deals with all effects occuring during changes in a world model. In a nutshell, the solution of the frame problem has to describe, what remains unchanged in the world model and what effects of a state change by operator application have to be noted.

Three general aspects of the frame problem are worth to be mentioned:

- the frame problem of the world model, i.e. which part of the real world has to be incorporated in the world model for achieving a correct model due to the associated operators
- the local frame problem, i.e. what is changed and what is stable after an operator application
- the frame problem of the user goal description, i.e. what if the user has specified the goal(s) incomplete.

MINOPTEROPS solves these frame problems. The solution of the first frame problem is very easy to obtain due to the great simplicity of the blocksworld. In a real world, it is extremely hard to solve this problem.

The second frame problem is solved due to the STRIPS operator philosophy in executing the "STRIPS-Assumption", i.e. the world remains unchanged until the explicit statements as part of the applicated operator what has to be deleted and what has to be added in the world model are executed. In this way, after each operator application the world model remains valid in terms of the operator execution. The frame problem due to user goals is very hard to solve, so that only hypotheses about the intention of the user goals can be implemented. MINOPTEROPS uses the "restore-hypothesis" /LOM/, i.e. restoring the initial state of the world in so far no explicit user goal is destructed. For implementation, rules are designed which formulate "synthetic user" goals in terms of the difference between the initial world state and the goal state.

As mentioned above, the implementation of the MEA algorithm in a forward chained language is cumbersome, but high valued in terms of studying every detail of the algorithm. The first implementation step of MEA-MINOPTEROPS was the design and implementation of a backward chaining mechanism by which a flexible goal directed MEA implementation was possible. An interesting detail during MEA algortihm implementation was the fact, that inside the execution of the backward algorithm very often the usage of the forward capabilities of OPS5 have been of a great benefit.

As discussed so far, only a symbolic plan in form of an operator sequence occurs at the end of the planning process. To reach and executionable plan, i.e. a plan which includes for example cartesian coordinates of the motions and action places, MINOPTEROPS transformed the symbolic plan in an execution module stepwise in direct executionable code with associated coordinate values. So the lowest level of autonomy can be reached using MINOPTEROPS as a task planning system for a robot.

5.2 COLAPS

Major drawbacks of MINOPTEROPS are the lack of capabilities for

- conjunctive goal handling
- optimal operator sequence generation
- explanation of the planning process
- improvement of efficiency.

Especially these facets are handled and solved by COLAPS /LOM/.

A significant increase of efficiency was reached during COLAPS development by procedural reimplementation of MINOPTEROPS in LISP /WIN1/. This cernel enables COLAPS to generate partial plans, i.e. operator sequences and associated world states, during possible solution tree generation phase. It should be noted, that this kind of improvement is a constant one. What is needed is a progressive improvement efficiency in face of the combinatoric explosion of the possible solution tree during complex planning tasks. To achieve a progressive behaviour, COLAPS incorporates filter functions which cut possible but at least inefficient or in the blocksworld physically impossible tasks. Reviewing figure 1, a simple function is

- it is inefficient to move a block, if there is another on it because it increases the number of actions.

Thus the possible solution tree can be cut by these heuristic worldmodel dependent functions. It can be seen from the principle tree structure, that such decisions in an early state of the solution process can reduce the problem solving effort dramatically.

Conjunctive goal handling and optimal operator sequence planning are handled from COLAPS by applying a heuristic search methode in combination with storing the possible world with each developed node. As mentioned above, the search procedure is defined by the branch-and-bound algorithm. The associated weighting factor is based on the number of activities necessary to reach the node under development.

Roughly speeking, the planning procedure works as follows. The user goals are analysed and transformed into a partial state of the goal world. During the second step a list of all possible core actions (i.e. in COLAPS the grasp-a-block action) in the initial state is generated and the actions are filtered by the "cut" functions mentioned above.

The remaining core actions are formulated as goals and are planned in a MINOPTEROPS manner. The fourth step evaluates the number of actions and assigns them to the associated world state. Thereafter, the world states are compared with the partial goal state and if no successful match can be reached, the node with the lowest weighting factor is developed by beginning the algorithm at the second step.

If the partial goal state is met by a developed node, all other possible branches will be developed further until every node has a weighting factor greater or equal to the solution node if not a shorter path to the solution is found. This procedure guarantees a minimal path of activity from the initial world state to the partial goal state.

From this, it can be seen how conjunctive goal handling is processed by COLAPS. The solution of this problem is based on the possible solution tree, containing a goal destructing world as well as a goal preserving world as a consequence of tree generation procedure. The efficiency criterion selects at least the goal preservation branch, because the goal destruction branch is in principal longer due to the number of goal destruction actions.

The explanation process can be handled in short, since from the tree structure of the solution graph all procedures mentioned above can be easily derived by associating each node of the possible world state a history file which contains parent action, child actions, filter decisions, state changes and weighting factor. Climbing up and down the history tree, all information about the planning process is available.

With regard to the user interface COLAPS stands in the history of SHRDLU, a blocksworld communication model, developed by Winograd /WINO/.

Basic idea of this concept is to provide a proper semantic to each syntactically correct user command including differently instantiated syntactic patterns. This allows a detailed and efficiently controlled dialogue between the planning system and the user.

Comparing MINOPTEROPS and COLAPS from the autonomy level, no increase of autonomy is reached but COLAPS is with regard to the requirements far more intelligent than MINOPTEROPS.

6. Conclusion

The AI techniques implemented in MINOPTEROPS and COLAPS are rather a very small facet of further applicable lowest autonomy level AI algorithms. But most of these algorithms deal also with an enhancement of the planners optimality, e.g. hierarchical/nonlinear planning /SAC2/, skeletal planning /FRI/, constraint directed planning /STEF1/STEF2/, etc. The associated systems all contain as a goal direction mechanism the MEA algorithm, a separate world model, more or less world specific heuristic filter functions and heuristic search techniques.

From the autonomy level point of view, there are some projects which are on a successful way to the highest level/ATK/.

However, basic intelligent techniques, necessary for that level, like representations and reasoning methods for temporal, spatial and commonsense aspects are still in the beginning of the development /WIL/, so that the question arises, how autonomous in the above defined sense these systems are. From the planning systems point of view, it is still a long way from now to a robot system on highest autonomy level.

References:

/ATK/ Atkinson, D.; Porta, H.
Autonomous Task Level Control of a Robot
Robotics and Expert Systems 1986

/BAF/ Barr, Avron; Feigenbaum, E.A.
The Handbook of Artificial Intelligence
Volume 1-3
Addision Wesley, 1982

/BRO/ Brownston, L.; Farrel, R.
Programming Expert Systems in OPS5
Addison Wesley, 1985

/FHN/ Fikes, R.E.; Hart, P.E.; Nilsson, N.J.
Learning and executing generalized robot plans
Art. Int. 3, 1972

/FIN/ Fikes, R.E.; Nilsson, N.J.
STRIPS: a new Approach to the Applicaiton of Theorem Proving and Problem Solving
Art. Int. 2, 1971

/FRI/ Friedland, P.
Knowledge-based experiment design in Molecular Genetics
Stanford University, Pept. No. Stan-CS-79-711, 1979

/LOM/ Lorenz, I.; Massau, C.;
Entwicklung wissensbasierter Plangenerierungssysteme für Industrieroboter in LISP und OPS5
Diploma Thesis, University of Bremen 1987

/NIL/ Nilsson, N.J.
Principles of Artificial Intelligence
Springer, 1982

/PEA/
Pearl, J.
Heuristics
Addision Wesley, 1984

/SAC1/ Sacerdoti, E.D.
Planning in a Hierarchy of Abstraction Spaces
IJCAI, 1975

/SAC2/ Sacerdoti, E.D.
A structure for Plans and Behavior
North-Holland, 1977

/STEEL1/ Steele, S.
The Butter and Bread of Planning
AI Review, 1, 1987

/STEF1/
Stefik, M.
Planning with Constraints
Art. Int., 16, 1981

/STEF2/ Stefik, M.
Planning and Metaplanning
Art. Int., 2, 1981

/WAL/ Waldinger, R.
Achieving several Goals simultaneously
Mach. Int. 8, 1977

/WIL/ Wilensky, R.
Planning and Understanding
Addison Wesley, 1983

/WINO/ Winograd, T.
Understanding Natural Language
Academic Press, 1972

/WIN1/ Winston, P.H.W.
Artificial Intelligence
Addison Wesley, 1984

/WIN2/ Winston, P.H.W.; Horn, B.
LISP
Addison Wesley, 1985

Collision Free Motion Planning for Robot-Manipulators

Ralf Heine, Thorsten Schnare

1. Abstract

An "intelligent" automatic system for planning safe trajectories for a computer controlled manipulator among obstacles is a key component of robot assembly operations. This paper describes an efficient algorithm, using configuration space, that finds collision-free paths for a manipulator with three revolute joints. Free space is described with a structure built up with "areas" and "cells" of three different sizes. These areas and cells are generated by a program written in MICROSOFT C. They are the database of the path planning program written in LPA PROLOG. Both programs are implemented on a PC AT 286. The generated paths are tested on a MITSUBISHI RM 501 manipulator (see fig. 1).

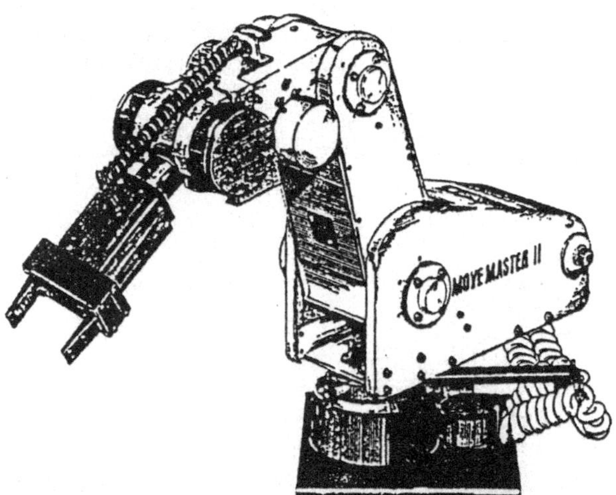

Fig. 1 : RM 501 manipulator

2. Introduction

Trajectory generation and path finding in the presence of obstacles are important research problems that have been attacked in various ways ([1],[2],[3],[4],[5],[6]). The motion-planning problem is to find a path from a specified starting robot configuration to a specified goal configuration that avoids collisions with a known set of stationary obstacles. Note that this problem is significantly different from, and quite a bit harder than, the collision detection problem: detecting whether a specified robot configuration or a path would cause a collision. Motion planning is also different from on-line obstacle avoidance: modifying a known robot path so as to avoid unforeseen obstacles.
A motion planning system has got the following inputs:

- a geometric and kinematic description of the robot
- a geometric description of its environment, made from optical sensor data or manual measurement, and
- the initial and final robot configurations.

The desired output is a trajectory for the manipulator.

If the obstacle configuration changes (e.g. further obstacles added) only the geometric input description of the environment has to be changed and the "intelligent" motion planning system is able to generate collision-free paths again. Such a system is much more flexible than a conventional system, which generates trajectories for one special order and environment. This conventional method is only useful in production lines without changes in the environment. An "intelligent" path-finding system is needed when handling with variant obstacle configurations.

3. General considerations

There are three classes of algorithms:

- hypothesis and test
- penalty functions
- free space algorithms

The first method proposes a path and tests it. When a collision is detected, a modification is made on rough approximations of the obstacle. The modification is quite rough and in a cluttered space a collision with another obstacle is likely. The penalty-functions encode the presence of an obstacle by a high value of the penalty-function in the proximity of an obstacle. These two methods use only local information, a more global view offer the free space methods which use a map representing the free configuration space in which the robot can move.

The "configuration" of a moving object is any set of parameters which completely specifies the position of every point on the object. The set of joint angles of a robot constitutes a configuration (robot coordinates representation). The cartesian parameters of the robot's end-effector (three for the position and three for the orientation) do not usually constitute a configuration because of the multiplicity of solutions to a robot's inverse kinematics. "Configuration space" is the space of configurations of a moving object. Thus, a robot's joint space is a configuration space. For manipulators with revolute joints the robot coordinates representation is profitable, because the constraints of the joints are automatically included. It is possible to map the obstacles in the robot's workspace into its configuration space ([6],[10]). These "configuration space obstacles" represent those configurations of the moving object that would cause collisions. "Free configuration space" is the complement of the configuration space obstacles.

The calculation of the free space has to be as much precise as possible. The characterization has to be compact to minimize the storage expence.

This paper describes a free space algorithm with global view. There are two main points:

1. Calculation of the whole free space and compact representation with a description using "areas" and "cells".
2. Using this database for searching a trajectory from the start to the goal point.

The program is written for a manipulator with revolute joints. The degrees of freedom (movable links) are constrained on three. The obstacles are described as cubes (see fig. 2).

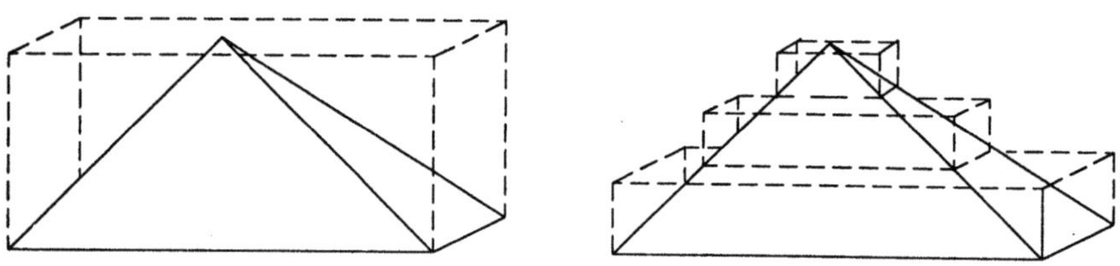

Fig. 2 : Two possible approximations for one obstacle by enwrapping cubes

Their geometrical positions are one of the algorithms inputs. Then a collision-free path from every chosen startpoint to every chosen goalpoint has to be calculated if at least one possible path exists.

The find path algorithm uses the advantages of logic programming. During the last step a safe trajectory described with a sequence of points in the configuration space is built up with the "cells", which connect the start- and the goal position. Every point in the configuration space describes the position of the three movable manipulator links. These points are sent to the manipulators "drive-unit" and the collision-free path is being executed.

4. Calculation of free space

The calculation of the free space should be simple and its characterization compact to reduce the storage expense. To make the calculation simple, first a so called "growing algorithm" is used ([1]). The manipulator and the obstacles are approximated by enwrapping cubes (see fig. 3a). The obstacles are enlarged in a manner allowing the manipulator to be represented by one-dimensional axes (see fig. 3b).

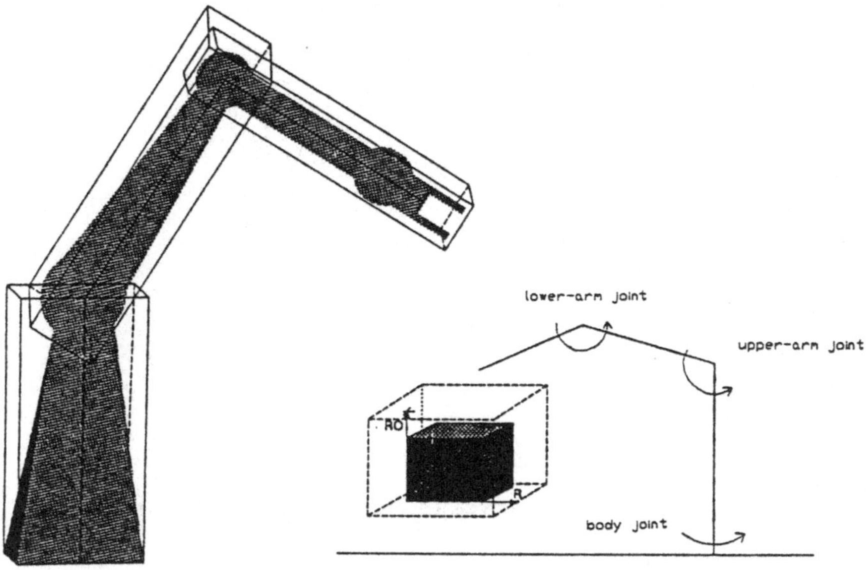

Fig. 3 : a) Manipulator approximation by enwrapping cubes

b) Obstacle growing and representing the manipulator by one-dimensional axes

To describe a manipulator with three joints the configuration space has to be three-dimensional. The free configuration space is calculated within two-degree steps in order to get a finite quantity of joint configurations. Making the resolution smaller than two degrees needs more computer resources. Making it more broad , it is likely to lose possible collision-free paths but there are fewer calculations to make. These calculations to get the description of free configuration space are simple but numerous. Regarding the MITSUBISHI RM501 manipulator, about 458,000 possible joint configurations exist using a calculation width of two degrees. It is very uneffective to check and save all 458,000 configurations one by one for collision. This was the reason to search for more effective methods.

The single collision-free configurations are combined to so called "cells". There are three different kinds of cells:

1. α-cells: They are described by a collision-free interval (α_u, α_o) for the manipulators body joint. Within the interval the manipulators motion is collision-free irrespective of the upper-arm and lower-arm joints. All possible configurations

of these two joints are allowed. There is no need to do more calculations (see fig. 4).

2. β-cells: They are described by a constant value α_w for the body joint and a collision-free intervall (β_u,β_o) for the upper-arm joint. Within the intervall the lower-arm joint may have any possible value (see fig. 5).

3. γ-cells: They are described by constant values α_w and β_w for the body and upper-arm joints and a collision-free intervall (γ_u,γ_o) for the lower-arm (see fig. 6).

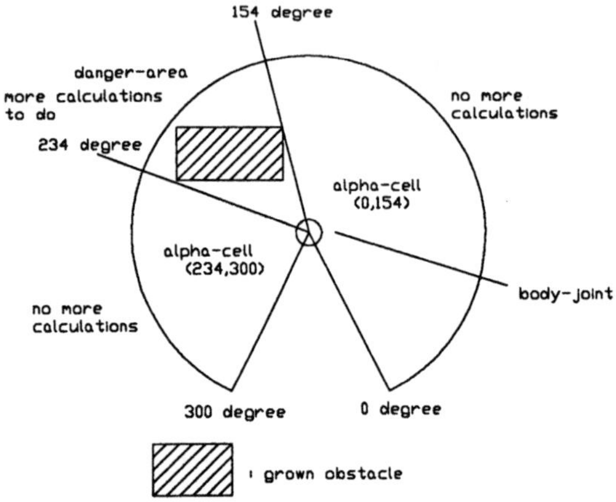

Fig. 4 : Plan view of working area with α-cell (example)

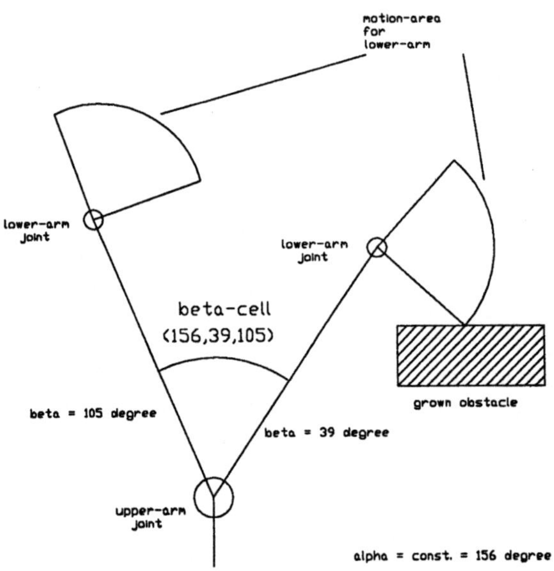

Fig. 5 : Side view with β-cell (example)

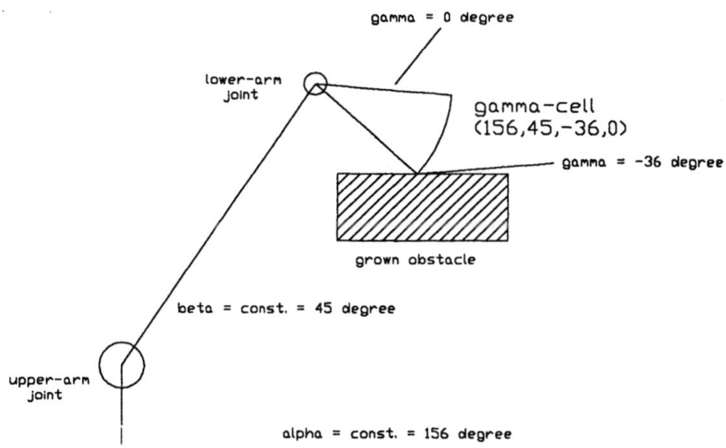

Fig. 6 : Side view with γ-cell (example)

The calculations to create these three different kinds of cells for representing the free configuration space is done by a program written in C-language. The free space calculation only has to be done once for one obstacle configuration. Then collision-free paths can be searched with this database until the obstacle configuration changes.

5. Searching for collision-free paths

The search for collision-free paths is done by a program implemented in PROLOG. The three different kinds of cells build the database for the search algorithm. It is implemented in PROLOG because of the powerful features like "backtracking" and "recursive calls" which enable to implement efficient search algorithms.

There are several search algorithms known in AI (Artificial Intelligence). A very efficient and directed algorithm is the so called A*-algorithm. It always finds an optimal path, when at least one path is possible.

But in spite of compact representation of free space, using the cell- method, the database is very large when representing complicated obstacle configurations. The A*-algorithm needs a long time to search the optimal path. This was the reason to generate a two-step search algorithm. Every step only uses a small quantity of the whole database. The database also consists of the three kinds of collision-free cells mentioned above. Connected β- and γ-cells for one constant α_w are combined to so called "areas" (see fig. 7a). The database is enlarged with the information, which areas are connected to each other (see fig. 7b).

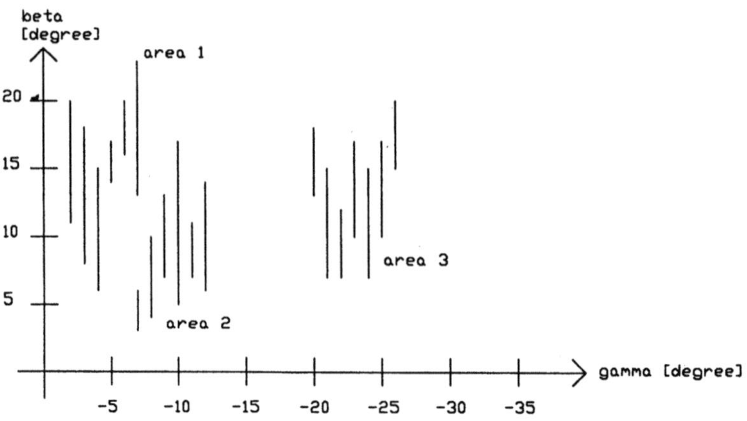

Fig. 7a : Three areas of connected cells for α = const.

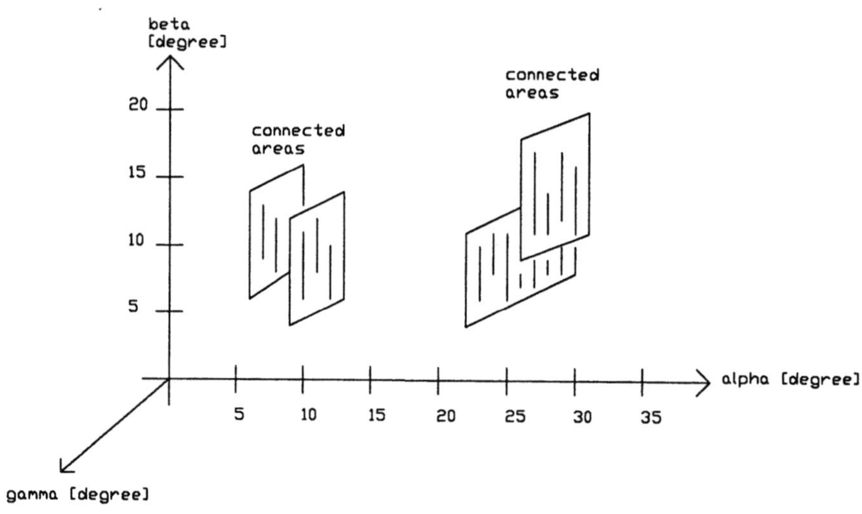

Fig. 7b : Connected areas

During the first step a collision-free path through the areas is searched from the start area, including the start position, to the goal area. In the second step a collision-free path through the simple cells is searched, going from area to area (see fig. 8).

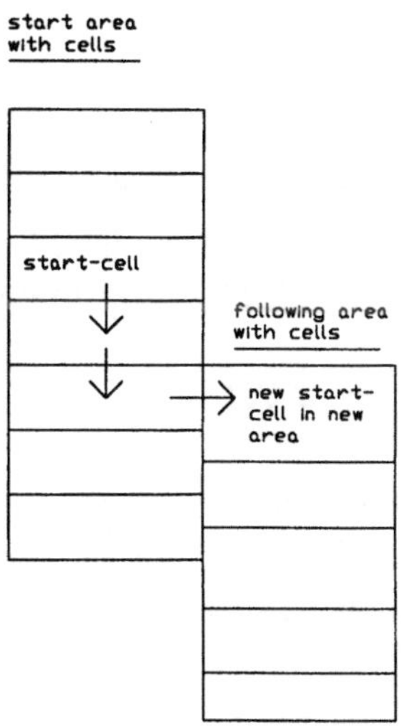

Fig. 8 : Searching a cell-path

The discovered trajectories are nearly optimal in many cases and even in complicated obstacle configurations a path through the cells is found within acceptable time. Finally the joint positions, describing a collision-free path for the manipulator are calculated. A further, so called "smoothing-step" removes redundant positions in order to get a smooth manipulator moving. The necessary positions are transmitted to the manipulators drive-unit. After programming the drive-unit the manipulators moving is started.

6. Conclusions

A practical algorithm is presented, which is able to generate collision-free paths among stationary obstacles. The obstacle configuration may change optionally. The free-space configuration method was used to represent the database for a two-step search algorithm, implemented in PROLOG (using an interpreter). The system is running on a 286 AT PC (10 MHZ). Calculation of free space needs about 1 minute even in cluttered obstacle configurations. Searching a collision-free path is done within several seconds up to about 1 minute depending on the length of the path and the obstacle configuration. The system was successfully tested with a MITSUBISHI RM 501 manipulator. It was demonstrated, that its able to do the work of planning collision-free path for a robot manipulator on a personal computer within acceptable time. Faster or future hardware and software developments (e.g. 386, 486 processor, PROLOG compiler) will decrease the amount of time for solving the find-path problem.

7. References

1. S. M. UDUPA *"Collision detection and avoidance in computer controlled manipulators"* Proc. 5th Int. Joint Conf. Artificial Intell. pp.737-748 1977
2. T. LOZANO PÉRES, R. A. WESLEY *"An algorithm for planning collision-free paths among polyhedral obstacles"* Communications of the ACM vol.32 pp.560-570
3. R. A. BROOKS *"Find-path for a PUMA-class robot"* MIT Artificial Intelligence Laboratory pp.40-44
4. R. A. BROOKS *"Planning collision-free motions for pick-and-place operations"* Intern. Journal of Rob. Research pp.19-44 1983
5. H. SAMET *"The quadtree and related hierarchical data structures"* ACM Computing surveys 16 pp.187-260 1984
6. B. FAVERJON *"Obstacle avoidance using an octree in the configuration space of a manipulator"* Proc. IEEE Int. Conf. Robotics pp.504-512 1984
7. T. LOZANO PÉRES *"Motion planning for simple robot manipulators"* Robotics Research Int. Symp. pp.133-140
8. T. LOZANO PÉRES *"Spatial planning: a configuration-space approach"* IEEE Trans. Comput. pp.108-120 1983
9. R. A. BROOKS *"Solving the findpath problem by good representation of free space"* IEEE Trans. Syst. Man Cyb. pp.190-197 1983
10. C. BAJAJ, M. S. KIM *"Generation of configuration space obstacles: The case of a moving sphere"* IEEE Journ. of Robotics and Automation vol.4 no.1 pp.94-99 1988

Register

Revised Index

Dobrivoje Popović (Ed.), **Analysis and Control of Industrial Processes**

Friedr. Vieweg & Sohn Verlagsgesellschaft mbH, Braunschweig 1991
ISBN 3-528-06340-8

GPSR Compliance

The European Union's (EU) General Product Safety Regulation (GPSR) is a set of rules that requires consumer products to be safe and our obligations to ensure this.

If you have any concerns about our products, you can contact us on ProductSafety@springernature.com

In case Publisher is established outside the EU, the EU authorized representative is:

Springer Nature Customer Service Center GmbH
Europaplatz 3
69115 Heidelberg, Germany

Batch number: 10584556

Printed by Printforce, the Netherlands